KAI DENHAM

THE FUTURE OF TECH UNDER THE TRUMP ADMINISTRATION

ARTIFICIAL INTELLIGENCE, CRYPTOCURRENCY AND TECHNOLOGY

The Future of Tech under the Trump Administration

Hamelin Bay Publishing
Hamelinbaypublishing.com.au

ISBN: 978-1-7643010-0-8
Book Cover Design and Interior Formatting by 100Covers.
First Edition 2025

Dedication and Acknowledgments

I am deeply thankful to the extraordinary people who stood beside me, offering their support and encouragement as this title came to life.

Dedication

This book is dedicated to Tess and Barbie

To Tess — my wife and steadfast partner in every chapter of life. Your love, patience, and unwavering belief in me have been the quiet strength beneath every step I've taken. You have encouraged me in ways too many to name, and this book carries your gentle, caring and empathetic fingerprints upon every page. Such a good listener!

To Barbie — my friend and my guiding light when the path ahead was unclear. You helped me find my way when I was lost, lending not only your wisdom but your faith in my voice. Your encouragement and belief have been the compass that kept me moving forward.

This work exists because of support and encouragement of these two wonderful ladies.

Acknowledgements

I acknowledge contribution and generous support of my family and friends — including Joanna, Ray, Marc, Susan, Greg and Graciella. Thank you for walking with me through the shaping of this manuscript. You read early drafts with patience and care, offering direct and honest feedback that challenged me to think more clearly, write more precisely, and refine my ideas into their strongest form.

Your positive suggestions were never just edits on a page; they were acts of generosity, given with the shared goal of making this work the best it could be. I am deeply grateful for the time you invested, the candour you brought, and the encouragement you offered at every stage. This book is sharper, clearer, and truer because of you.

With deep gratitude to all who have supported me, I thank you for your support.

Kai

Table Of Contents

Introduction

We live in an era where technology pervades every facet of our lives, the boundaries between personal choices and governmental oversight blur into a tangled web. Your cell phone seems to know you better than you know yourself. It already knows what you want to buy, where you want to go, and what will make you laugh. It also shows you people expressing the same ideas you have about all sorts of topics. It is much easier to make choices because you quickly reach the information you need to be certain. However, is it truly easier or are you losing freedom by only focusing on the options already aligned to a single idea?

In our current world, your digital footprint is not only monitored but also strategically maneuvered by those in power. What would happen if policymakers wielded their positions to shape narratives, control information, and influence decisions that you thought were yours alone?

President Donald Trump took office in the White House for his second term in January 2025 and has implemented innovative and disruptive changes since his first days in power. He can be a polarizing figure who has once again stirred the waters with sweeping executive orders that redefined America's technological landscape. With a single stroke of his pen, policies from the previous administration are retracted, all in favor of promoting what he defines as "American innovation." This isn't just about ideological shifts; it contributes to creating a new reality where artificial intelligence (AI) and cryptocurrencies become pillars of governance rather than mere subjects of debate.

One of the latest developments is, the formation of the President's Council of Advisors on Science and Technology (PCAST), which includes influential tech executives and outspoken advocates for AI and cryptocurrency. It inevitably raises questions about how these inter-twined connections between political power and tech CEOs will reshape policy and impact free speech. These two spheres with different approaches and objectives are supposed to work together to steer the country away from what they describe as Big Tech bias. But what does this truly mean for us as citizens navigating this rapidly changing global landscape?

Each move, each proposal, carries profound implications for how we understand privacy, information access, and even fairness in our societies. Freedom of thought, speech, and free will are the backbone of contemporary political systems and have been the most prized values since the earliest civilizations. It is the basis of human dignity. However, power has always found a way to influence people's decisions either to sustain the status quo or to turn the course of events to the group's benefit. As a counterpart, citizens have created ways to resist, voice their needs, and influence political power.

The theater in ancient Athens, the Colosseum in the Roman Empire, later the printing press, and much more recently, mass media served as a two-way path for political power to meet people's expectations. There has always been an endless interplay between both sectors, but it has never been a fight on equal terms. There is asymmetry, and the political power is always on top. Even though citizens keep for themselves the right to choose, those choices are framed by social discourse and propaganda.

These processes have reached unprecedented scopes. At present, big data, the omnipresence of social media, and AI acting behind the curtain are flooding every aspect of daily life in the most subtle ways. We are still figuring out how much it can impact our lives. How much can political power manipulate these new information and communication technologies that have even reached the financial and economic system?

We begin with a deep dive into how AI and digital tools already pervade our homes, wallets, and workplaces. From there, we explore the historical roots and present-day effects of regulation and deregulation, tracing how executive actions, like those during the Trump administration, ripple through every sector from fintech to national security.

Later chapters engage with hot-button topics: the rise of digitally driven movements, the ethical dilemmas of campaign technology, the proliferation of cryptocurrency, and the quiet evolution of surveillance and military norms. Real world examples anchor each topic, while guided reflection prompts invite you to pause and consider what these changes mean for your role as a citizen. The goal isn't to overwhelm with jargon, but to make even the most complex subjects approachable, relevant, and alive.

The stakes could hardly be higher. Digital transformation offers extraordinary promise: more efficient services, increased access, and new forms of social connection. Yet every advance brings trade-offs, and there are no simple binaries between progress and peril.

Throughout, this book resists the temptation to offer quick fixes or lean into alarmism. Instead, it creates genuine dilemmas—how do we balance innovation and accountability, personal privacy and collective safety, free speech and responsible moderation? Rather than pushing a single viewpoint, the text invites you to wrestle with complexity, recognizing that sustainable solutions require listening and adaptation.

Besides the impact on sensitive information, we delve further into the unfolding and fast changing environment of cryptocurrencies, the digital assets that are redefining economic landscapes. Policymakers around the globe are grappling with the dual promise and peril presented by them. Cryptocurrencies are opportunities for financial inclusivity, but they also hold the potential for misuse and regulatory evasion. Political figures have either embraced or resisted this shift following opposing interests and motivations.

On our end, we want to know the reasons behind the decisions and consider the ethical controversies at play. Moreover, we aim to understand the direct impact upon the global market and our pockets.

Ultimately, this book is grounded in optimism, that informed and engaged people can shape the trajectory of democracy in the digital age. Civic trust, resilient institutions, and ethical choices remain within reach if we approach the challenges ahead with openness and resolve. Whether it's embracing new opportunities in artificial intelligence or scrutinizing policy decisions, your curiosity, skepticism, and willingness to engage aren't just valuable, they are essential for building a fairer, stronger society.

A few years ago, a friend confided her growing discomfort with the 'personalization' she saw online, her feeds seemed to know more about her thoughts than she did. She'd always considered herself vigilant, but realized that despite filters and settings, it was often difficult to tell which content came from thoughtful sources and which had been engineered to provoke, distract, or sway. Her story reflects the lived experience of millions, navigating a world where information feels both empowering and exhausting and where the lines between engagement and manipulation are increasingly hard to see.

With this book, I aim to offer both context and companionship to those seeking a deeper, more discerning understanding of the future evolving digital reality under the influence of the Trump Administration.

Chapter 1

Understanding the Technological Landscape

The frantic rhythm of our lives and the constant and rapid evolution of technology create a sense of things as if they have always been here. However, in the past two decades, everything has drastically changed.

Now, with the return of Donald Trump to the presidency in 2025, the intersection of politics and technology enters a new phase. His administration brings a distinct, often polarizing approach to regulation, innovation, and control. While many details of his second-term agenda are still unfolding, it's clear that his stance on Big Tech, AI, data privacy, and decentralization will diverge sharply from previous years. Before diving into the implications of his policies and the legal shifts ahead, it's important to first understand the current technological landscape—the foundation on which these political decisions will build.

A few decades ago, we imagined a future with floating cities, flying cars, and vacations on Mars. While some of that is actually happening, with a few fortunate individuals taking trips to outer space, these types of opportunities are limited to a small group. Instead, technological advancement has made incredible, widespread progress in the past couple of decades, reaching heights we never anticipated and influencing areas of life we hadn't considered.

Telephones and the Internet were remarkable inventions that transformed communication, but we have advanced so much further with augmented reality (AR), virtual reality (VR), and artificial intelligence (AI). We aren't just interacting with others without the limitations of distance; we have also transcended the boundaries of time and reality, engaging with holograms and bots. Since the invention of money, it would have been hard to imagine profiting from something solely based on mutual trust and hardly any physical support. However, in simple terms, that's the cryptocurrency market that many of us are enthusiastically delving into.

The world has quickly become a fascinating and challenging science fiction plateau, and we weren't given the script in advance. While we try to remember the dozens of passwords and user names to navigate this new world, a reduced group of powerful people control the backing board. Some of them are tech industry managers and developers who are literally crafting the stage and shaping how the characters should behave; others are the policymakers with the power to set the rules of the game.

From the subtle tailoring of your social media feed to the thermostat that remembers your evening preferences and the health app monitoring your heart rate, artificial intelligence constantly shapes ordinary moments. While AI brings unmatched convenience, it also raises pressing questions about how much we trust machines with our privacy. Ongoing debates about data collection, opt-in choices, and the right to control personal information remain closely tied to broader questions about the future direction of regulation and policy, as explored further in this text. (Fazlioglu, 2024; Granados, 2024).

Today's tech landscape is the playing field. Fintech, AI, big data, and crypto aren't just flashy terms thrown around in headlines; they're the tools actively reshaping how we handle money, make decisions, and even relate to one another. These forces are breaking down old systems, opening up new possibilities, and raising big questions along the way. Even if it's it's how we bank, trade, predict trends, or assign value, these technologies are baked into our everyday lives. Now, let's take a look at how these innovations are coming together to shake up financial markets, change business models, and challenge the rules as we know them.

The Evolving Landscape of Internet and Social Media Use

Over the past decade, daily life has become increasingly intertwined with the internet. By 2023, nearly half of all young people reported being "always online," and smartphone ownership had become nearly universal among younger generations. The COVID-19 pandemic only deepened this shift, pushing work, education, and social interaction further into the digital sphere and making even the simplest of tasks reliant on technology (Firth et al., 2024).

The Rise of Streaming and Short-Form Content

Entertainment habits have shifted rapidly as traditional media formats give way to digital options. Music streaming platforms and podcasts now compete directly with radio, with users spending an average of 100 minutes daily streaming music and nearly an hour listening to podcasts. Video streaming has also taken center stage, with YouTube emerging as the most widely used social media platform. Short-form videos—popularized by TikTok, Instagram Reels, and YouTube Shorts—have transformed content consumption habits, favoring quick bursts of entertainment over longer, conventional formats.

Social Media: From Habit to Habitat

Social media is no longer a side activity; for many, it's the primary digital experience. Working-age adults now spend an average of 2.5 hours a day on various platforms, making up nearly 40% of their total internet usage. Among teens, the number who consider themselves "constantly online" has nearly doubled since 2015. These patterns have drawn increasing scrutiny from health professionals, researchers, and governments concerned about the psychological and social implications of such high engagement levels.

Building Connection Across Borders

Despite concerns, social media remains a powerful force for connection. Platforms like Facebook, X, LinkedIn, and Reddit enable users to form communities around shared interests, ideologies, or life experiences.

Whether it's climate activism, mental health awareness, or niche hobbies, online communities often transcend national and cultural borders. For instance, Facebook groups offer safe spaces for like-minded people to bond, organize, and mobilize, fostering a sense of belonging and support that can be hard to find offline.

Support in the Digital Age

For those facing life's challenges—whether it's parenting, chronic illness, mental health, or addiction—social platforms offer more than just content; they offer connection. Online forums and anonymous support groups provide judgment-free zones to share experiences and gain emotional strength. Real-time features like WhatsApp calls, Zoom meetings, and live streams on Instagram and Facebook bring immediacy to these interactions, helping people celebrate milestones, collaborate professionally, or simply stay emotionally close despite physical distance.

The Cost of Constant Connection

However, this hyper-connectivity comes at a price. The nonstop stream of updates, ads, and viral content often leads to overstimulation, displacing meaningful communication with mindless scrolling. Many users report feeling trapped in a cycle of passive consumption that leaves them feeling drained. In response, people are increasingly turning to self-regulation tactics—muting notifications, uninstalling apps, planning screen-free time—in an effort to reclaim their attention and mental space. Yet even with these strategies, disconnecting can feel impossible during moments of emotional vulnerability (Firth et al., 2024).

Rethinking the Metrics of Impact

Researchers Friedman and Delerme (2025) caution against measuring the impact of social media solely by time spent online or user demographics.

Instead, the type of engagement matters. While over 90% of adolescents report at least one negative consequence from tech use, many also cite benefits such as easier access to education, career opportunities, and emotional support. The distinction between active engagement—like posting and messaging—and passive use, such as scrolling or lurking, is still being studied, and findings remain inconclusive.

A Double-Edged Tool

Ultimately, social media is a double-edged tool—capable of amplifying connection, knowledge, and empowerment, while also introducing new challenges to mental health and emotional well-being. The same platforms that bring people together across continents can also contribute to stress, distraction, and disconnection from the present. Moving forward, the key will be not only to expand access and innovation, but to support healthier, more intentional digital habits for all users.

Artificial Intelligence and Its Applications

Among today's tech advances, artificial intelligence stands out—not just for what it can do, but for the ethical conversations it sparks. It's one of those things that feels futuristic and familiar all at once. AI systems are designed to mimic human thinking, but with the capacity to process massive amounts of data at lightning speed. That means faster predictions, smarter automation, and a growing presence in just about every industry.

At its core, AI is powered by algorithms that learn from data. These systems can now handle tasks we once thought were uniquely human—recognizing speech, translating languages, making decisions, even interpreting visual information. Because of this, we see AI popping up in places as varied as medical diagnostics and financial forecasting. The common thread? Speed, scale, and the ability to identify patterns we'd likely miss.

What's especially interesting is how often we interact with AI without even realizing it. It's behind the curtain, woven into the tools we use every day. Ask Siri to set a reminder, or let Google Assistant find your next meeting location—AI is quietly working in the background. The same goes for those eerily accurate shopping recommendations or the curated playlists that seem to "know" your mood. Whether it's Amazon suggesting your next purchase or Netflix lining up your weekend binge, AI is constantly learning from your choices to offer something personalized.

In the business world, AI has become a game-changer. Automation cuts down on repetitive tasks and reduces the risk of human error, all while increasing efficiency. Take customer service, for example—AI-powered chatbots can answer questions around the clock, often resolving issues faster than human reps. That frees up employees to focus on the more complex stuff and helps companies deliver better service overall. On the strategy side, AI helps businesses dive deep into customer behavior, forecast trends, and make smarter decisions based on real-time insights. It's not just about speed—it's about clarity and staying competitive in a rapidly changing market.

But while the benefits are real, so are the concerns. A closer look reveals that a small group of tech giants like Google, Microsoft, and Amazon, dominate the AI landscape. With their access to vast datasets and immense computing power, these companies are setting the pace for innovation. But their influence also raises questions about market fairness. When a few players hold most of the cards, the risk of an AI oligopoly becomes very real potentially limiting competition, creativity, and even democratic access to powerful tools.

Outside the business arena, AI is also changing public services and healthcare. In hospitals, it helps doctors diagnose diseases faster, tailor treatments to individual patients, and predict outcomes with surprising accuracy. In cities, AI is being used to improve infrastructure, manage traffic, and even reduce energy waste—contributing to broader sustainability goals in the process. These examples show that AI is reshaping how we care for people and the planet.

Emerging Challenges

Beyond market-related concerns, one of the most pressing issues is the potential for bias within AI systems. Bias can emerge from skewed training data or flawed algorithms, leading to unfair treatment of individuals based on race, gender, or socioeconomic status. These scenarios highlight the urgent need for ethical standards and transparent practices in AI development and deployment. Systematic checks must be implemented to identify and reduce bias, ensuring that AI operates fairly across different societal contexts.

AI also raises serious privacy concerns, particularly around the collection, storage, and use of personal data. Key questions arise: Who is gathering this information? How can it be used to influence our decisions? And what impact might that have—especially on politics? As previously noted, only a small group of AI providers currently control this type of sensitive data, which gives them immense power in shaping user behavior and public discourse.

Tech companies and AI providers often advocate for looser regulations, arguing that fewer restrictions promote innovation and allow for faster progress. From their perspective, a lighter regulatory environment encourages competitiveness, drives economic growth, and speeds up the rollout of AI solutions across different sectors.

However, this push for deregulation has sparked concerns among policymakers and the public. Critics warn that too little oversight could result in unintended consequences—such as bias, misuse of personal data, and the worsening of existing inequalities. As a result, the debate continues over how to strike the right balance between encouraging innovation and ensuring AI is developed and used ethically and responsibly.

Impact on Jobs

Another major concern is AI's role in job displacement. Early in Trump's new term, his administration collaborated with Elon Musk to establish the Department of Government Efficiency (DOGE), a move intended to reduce federal spending (Kinnard, 2025). The initiative marked a significant shift in federal workforce management and led to thousands of job losses. While some referred to the early collaboration between Trump and Musk as a "bromance," the partnership ultimately ended following backlash over the aggressive use of AI to replace human labor.

In the initial months of the administration, thousands of federal employees, particularly probationary staff, were let go. The White House encouraged agencies to explore technology-based solutions for streamlining their workforce. Though appealing to a tech-forward mindset, this shift had real consequences for workers, including those in critical roles at the Department of Veterans Affairs and the Department of Defense.

The introduction of DOGE and AI-based tools allowed agencies to automate workforce evaluations by analyzing employee performance data and departmental needs. These systems influenced decisions about who stayed and who was dismissed—placing unprecedented weight on algorithmic assessments to predict workforce trends and "optimize" staffing levels.

Basics of Fintech and Its Impact

Fintech—short for financial technology—marks a significant evolution in the financial sector. Fueled by rapid innovation, it's transforming the way individuals and institutions manage money. At its core, fintech encompasses a wide range of tools and platforms designed to enhance and automate financial services, often bypassing traditional banking systems in favor of greater speed, access, and personalization. As smartphone and internet access continue to expand globally, more people are engaging with fintech solutions that streamline interactions between consumers and financial institutions.

Mobile banking, for example, has made everyday financial tasks quicker and more convenient. Gone are the days of waiting in line at a physical branch. Users now check balances at midnight, move money between accounts in seconds, and even deposit checks by simply taking a photo. This always-on access puts greater control in the hands of users—but also creates new vulnerabilities. Cybercriminals increasingly target mobile apps to steal credentials or intercept data. In response, banks are deploying biometric authentication like fingerprint and facial recognition, encrypting user data, and monitoring activity in real time. These safeguards rely heavily on AI algorithms that adapt to user behavior, flag suspicious activity, and continuously improve fraud detection (Xu et al., 2025).

Peer-to-peer (P2P) payment platforms have added a social and intuitive layer to how we move money. Apps like Venmo, Zelle, and Cash App make it simple to split dinner bills, share rent, or reimburse friends—with transactions often accompanied by emojis or casual messages. Some platforms even allow groups to track shared expenses, easing the awkwardness of who owes what. But the convenience comes with risks: scammers may exploit the system, or unauthorized transfers may occur if a device is compromised. To mitigate this, regulators require platforms to implement identity checks and transaction monitoring. Companies must strike a careful balance—complying with anti-money laundering laws while preserving the simplicity and speed that make these apps so appealing (Chand et al., 2025).

Emergence and Growth of the Fintech Sector

The personal finance landscape has been simplified through fintech. The financial crisis of 2008 catalyzed rapid fintech adoption. This global economic downturn was triggered by the collapse of the housing bubble in the U.S. Risky mortgage-backed securities led to massive bank failures, causing widespread panic, significant job losses, and government bailouts—ultimately reshaping financial regulations and consumer trust in financial institutions.

Public trust in conventional banks diminished, leading to a demand for more transparent, efficient, and customer-centric solutions. Today, physical bank branches are becoming increasingly rare, replaced by a wide variety of money management apps and digital wallets on our smartphones. This shift signals a new era in personal finance.

Expansion and Market Momentum

Fintech companies span various types, including start-ups, growth companies, banks, non-bank financial institutions, and cross-sector firms. These include peer-to-peer payment services as noted previously, automated portfolio managers, and trading applications such as Robinhood and Coinbase. The fintech sector began gaining prominence around 2010, especially in the payments space, with innovations like Square enabling small businesses to accept credit cards via mobile devices. By 2021, nearly half of U.S. consumers utilized fintech products, illustrating its rapid growth (McKinsey & Company, 2024).

However, after an explosive rise in venture capital funding—from $19.4 billion in 2015 to $33.3 billion in 2020—2022 saw a market correction. In response, fintech companies began redirecting their focus toward sustainable, profitable growth. Traditional banks, in turn, started creating their own fintech-informed digital products and services, intensifying competition for client deposits. As fintech continues to evolve, its convenience and accessibility are reshaping the future of financial services, influencing how users store, save, borrow, invest, transfer, pay, and protect their money across the globe.

Cryptocurrency and Credit Innovation

The rise of cryptocurrency marks another wave of transformation in financial services. Initially dismissed as speculative, digital coins such as Bitcoin, Ethereum and Solana now double as alternative currencies and investment assets. Several retailers allow shoppers to pay with cryptocurrency either in person or online, while international firms use stablecoins for cross-border transactions to avoid costly currency conversions or delays. Cryptocurrency wallets can store funds securely and enable anyone with an internet connection to participate in the global economy. Yet, price volatility remains a significant barrier—values can swing dramatically in short periods, making budgeting and long-term use risky for everyday consumers. Regulation is slowly catching up, with some governments recognizing digital coins as legal tender and others developing frameworks to address risks of fraud, money laundering, and market manipulation alongside innovation benefits (Xu et al., 2025).

Data-driven lending platforms have changed how people and businesses access credit. Rather than depending solely on traditional credit scores, new lenders analyze a broader set of information to assess risk and offer loans. This can include utility bill payments, e-commerce purchasing patterns, rental records, or business performance metrics gathered from digital sources. Sophisticated algorithms review this data and predict whether an applicant is likely to repay a loan, enabling approval decisions in minutes rather than days. Entrepreneurs and individuals with limited or no formal credit history now have more chances to qualify for financing. But the extensive data collection required by these systems raises questions about privacy, consent, and potential bias. Companies must be transparent about what information they gather and how it influences decisions, as well as safeguard personal details against misuse or breaches (Xu et al., 2025).

Everyday Impact and Ongoing Questions

Each of these fintech innovations directly affects daily life. Consumers have become accustomed to managing money without ever entering a physical bank, splitting restaurant tabs at the table, or borrowing funds without paperwork. Behind these changes, artificial intelligence makes banking smarter and more efficient, setting the stage for how people communicate and interact financially. Privacy remains a thread running through all advances—balancing convenience and protection shapes not just current applications but the entire future trajectory of technological change. By reimagining core financial activities, fintech lays groundwork for an even deeper transformation yet to come, connecting finance, communication, and technology in unprecedented ways.

Fintech Past and Present

From a historical perspective, we can draw a parallel with prior regulatory efforts to balance innovation and consumer protection. For instance, regulations introduced in response to major financial disruptions often aim to ensure fair competition and secure consumer interests. These regulatory frameworks evolve alongside fintech advancements to address emerging concerns without hindering technological progress. Analyzing past regulation examples helps anticipate how current trends might unfold and identify potential intervention points for policymakers to safeguard economic stability (Del Sarto et al., 2025).

Looking into the future, fintech's trajectory suggests continued evolution, tightly interwoven with AI and other advanced technologies. One anticipated trend involves utilizing AI capabilities to further personalize consumer experiences. AI-driven insights could offer tailored recommendations based on user behavior and preferences, thereby enriching financial decision-making processes. Additionally, advances in big data analytics promise to enhance risk assessment and fraud detection mechanisms, providing both security and efficiency in financial transactions.

However, as fintech becomes more integrated into everyday life, regulatory adaptation remains paramount. Policymakers are tasked with crafting frameworks that account for technological change while safeguarding individual rights. This includes addressing privacy concerns and ensuring equitable access to emerging fintech solutions. The challenge lies in striking a balance between encouraging innovation and mitigating potential risks associated with rapid digital transformation.

Role of Cryptocurrencies in the Financial Market

The world of cryptocurrencies represents a turning tide in how we understand and interact with money. Originating from the innovative idea of Bitcoin, cryptocurrencies emerged to challenge traditional monetary systems that have been in place for centuries. At the heart of Bitcoin and other digital currencies is cryptography, a tool used to secure transactions and control the creation of additional units. This technology offers a decentralized model, disrupting conventional financial structures by enabling direct transactions between individuals without the need for an intermediary like a bank. This decentralization opens up new economic opportunities, allowing for more inclusive financial systems across the globe.

Blockchain technology underlies these cryptocurrencies, providing a foundation defined by transparency and security. Think of blockchain as a public ledger where every transaction made in a cryptocurrency network is recorded in a secure and immutable manner. This characteristic not only ensures that transactions are safe but also promotes user sovereignty over wealth. Users can conduct peer-to-peer transactions seamlessly, with no need for third-party validation, thus maintaining privacy and autonomy. This revolutionary use of blockchain technology allows individuals to send money worldwide quickly and securely, potentially replacing old banking technologies as more people gain access to smartphones.

Key players in the cryptocurrency markets wield substantial influence, shaping how these digital assets are perceived and regulated globally. Their motivations range from profit-making to establishing cryptocurrency as a legitimate and integral part of modern finance. However, their involvement raises questions about potential risks, where few large entities could dominate the market, reducing competition and innovation.

This scenario necessitates careful regulation, something observed when governments around the world, including the U.S. Commodity Futures Trading Commission, began regulating Bitcoin as a commodity (Javaid et al., 2022). The main purpose of regulations is to strike a balance between innovation and protection for investors from fraud and volatility.

Market dynamics in the cryptocurrency realm are notably volatile, characterized by rapid price fluctuations that pose both risks and opportunities for investors. This volatility is heavily influenced by numerous factors, such as regulatory news, adoption rates, and technological developments.

Cryptocurrency volatility presents significant risks, particularly with speculative assets like meme coins. These coins, often lacking real utility, can experience drastic price fluctuations within short periods. A concerning tactic seen in this space is the "rug pull" strategy, where developers abruptly withdraw liquidity, leaving investors with worthless assets. The case of $LIBRA in Argentina highlights these issues, as President Javier Milei promoted the meme coin, which surged to a $4 billion valuation after launch. However, as early investors cashed out, the price plummeted, leading to accusations of fraud and prompting an investigation into Milei's involvement in the project (Suderman & Debre, 2025).

Exploring the stakeholders' landscape reveals a spectrum of participants, from individual investors and tech companies to financial institutions and governments. Each has unique interests and stakes in the evolution of cryptocurrencies. Regulations play a crucial role in shaping stakeholder interactions, as differing rules across countries can affect strategies and market access. Therefore, understanding how political involvement and international regulatory landscapes impact the market is vital for anyone engaged in cryptocurrency activities. However, there is still a broad legal vacuum that challenges the reliability of the operations and the stability of the market.

The complexity of cryptocurrencies extends beyond their technical framework to include significant economic implications. By challenging the conventional monetary system, they offer alternative solutions that may address inefficiencies in cross-border transactions and provide financial services to unbanked populations.

As technology continues to advance at an astonishing pace, the need for effective cybersecurity practices has become paramount in protecting both technological innovations and their users. Cybersecurity can be defined as the set of practices designed to safeguard systems, networks, and data from a variety of cyber threats. This foundational understanding underpins trust in our increasingly digital environments. A robust cybersecurity framework protects sensitive information and ensures the integrity of individuals and organizations.

Bringing It All Together

- Throughout history, every major technological revolution has disrupted traditional social structures, often triggering debate, controversy, and adaptation.
- Unlike past eras, political power is now playing an increasingly active role in promoting and shaping technological advancement.
- As we move forward, ongoing dialogue and thoughtful policy-making will be essential to:
    - Maximize the benefits of innovation,
    - Mitigate associated risks,
    - And ensure that progress aligns with core societal values.

- In the United States, President Trump has adopted a proactive stance on technology, pledging to position the country as a global tech leader.
- Shortly after taking office, his administration issued executive orders aimed at:
    - ▷ Reforming regulations around artificial intelligence,
    - ▷ Encouraging growth in the cryptocurrency and digital asset space,
    - ▷ And promoting fintech as a key driver of economic dynamism.
- However, the long-term consequences of these policy shifts remain to be seen, leaving open important questions about their broader impact on society and the economy.

Chapter 2

Global and Historical Context and Policy Shifts

Technology has evolved alongside humanity, shaping every social structure we've built. But the pace and depth of change have accelerated dramatically since the Industrial Revolution of the late 18th century. According to Schwab (2016), we are now in the midst of a Fourth Industrial Revolution, characterized by three defining factors: velocity, scope, and systems impact. Technological breakthroughs today are occurring at unprecedented speed, evolving exponentially rather than linearly. This revolution is disrupting nearly every global industry, reshaping how we produce, manage, and govern across all sectors.

Throughout history, societies and governments have adapted in response to technological disruption. In the 1990s, landmark legislation was introduced to regulate the newly public Internet. Designed to foster competition and lower entry barriers in the telecommunications market, it also sparked new challenges. By the 2000s, the debate around net neutrality emerged—raising critical questions about equitable access and the potential for discriminatory practices by Internet service providers. In recent years, privacy laws have taken center stage, reflecting growing concerns over consumer data protection in an era of constant digital surveillance and commercialization.

This chapter explores how regulatory approaches continue to evolve, shaping the future of innovation. It examines the effects of recent deregulation measures, comparing how various countries approach technology governance through different laws and frameworks. By looking at the legacy of early legislation, the complexities of deregulation, and the diverse models of global policy, readers gain a deeper understanding of the ongoing tension between innovation and oversight. Policy choices are not made in isolation—they ripple across industries and societies, influencing both the pace and the direction of technological progress.

This discussion also sets the stage for why AI, cryptocurrency, and fintech became central to President Trump's early policy agenda during his second term. These technologies not only represent cutting-edge innovation, but also stand at the crossroads of politics, economics, and regulation—making them a natural starting point for examining how government power and technological ambition collide.

Global Comparisons in Technology Policy

Europe's General Data Protection Regulation (GDPR), enacted in 2018, remains the gold standard for global data privacy. It requires companies to obtain explicit user consent before collecting or processing personal information, notify users of data breaches within seventy-two hours, and offer simple tools for accessing, correcting, or deleting personal data. Key provisions include the right to data portability and the obligation for data controllers to implement strong security measures. These rules have pushed many non-European companies—particularly U.S. tech firms—to restructure platforms, update privacy policies, and limit certain data-driven features. Compliance has come at a high cost, with some companies choosing to pull products from the EU market entirely. Looking ahead, the EU's proposed AI Act, expected to be finalized in 2025, introduces risk-based classifications for AI applications and mandates transparency and human oversight for high-risk systems such as biometric surveillance and financial decision-making tools. Europe's regulatory approach is wide-reaching and often unforgiving, especially for companies with lax standards (IEEE Digital Privacy, 2025; Gcore, 2024).

The ripple effects of GDPR extend well beyond Europe's borders. Even American companies operating in largely deregulated environments must comply with GDPR if they process data from EU citizens. This extraterritorial reach demonstrates how a single region's legal standards can influence global practices—creating both friction and forward movement for tech innovators worldwide. As discussed in Chapter 1, the Trump administration favored aggressive deregulation at the federal level, prompting states like California to step in with their own measures. The California Consumer Privacy Act (CCPA) borrowed heavily from GDPR, highlighting the fragmented regulatory landscape U.S. companies must navigate.

Asia takes a distinctly different approach, blending rapid innovation with a range of regulatory models. In China, some of the world's strictest data localization laws require that companies store data within national borders and submit to rigorous government audits. The Personal Information Protection Law (PIPL) tightly aligns data governance with state objectives—prioritizing national security, platform control, and technological self-sufficiency. Cross-border data transfers face high scrutiny. By contrast, Singapore embraces a public-private partnership model, using frameworks like its Model AI Governance Framework and regulatory sandboxes to encourage innovation while maintaining oversight. South Korea has introduced its AI Industry Promotion Act, focusing on transparency and data localization without the sweeping control mechanisms seen in China.

Meanwhile, countries like Japan combine voluntary "soft law" codes with gradually evolving regulations that aim to preserve both privacy and innovation. India is moving in a similar direction with its Digital Personal Data Protection Bill, which seeks to harmonize consumer rights with the need for growth in its digital economy. These variations reflect different national priorities: China's top-down strategy contrasts with Singapore's collaborative experiments and Japan's incremental policy development. Across the Asia-Pacific region, however, shared themes emerge—especially around transparency and data sovereignty—even as countries take divergent paths. Overall, Asian digital policies tend to favor business flexibility and social cohesion, often avoiding the adversarial, enforcement-heavy approach seen in the West (Gcore, 2024; IEEE Digital Privacy, 2025).

Amid this global patchwork of approaches, transnational data policy remains one of the most difficult challenges. Jurisdictional conflicts frequently arise—whether from GDPR's restrictions on international data transfers or China's rigid localization demands—clashing with the inherently borderless nature of cloud infrastructure and global commerce.

U.S. companies are increasingly caught in the middle, trying to comply with overlapping or even contradictory requirements as they move data across Europe, China, and emerging markets. Regional efforts such as the APEC Cross-Border Privacy Rules system and the African Union Convention on Cybersecurity reflect attempts to coordinate policies, but true global interoperability remains elusive. Inconsistent enforcement and a lack of universal norms continue to complicate compliance, highlighting the need for ongoing cross-border dialogue and regulatory cooperation (IEEE Digital Privacy, 2025).

Historical Examples of Technology Policy

The history of technology legislation in the United States reveals a series of pivotal milestones that have shaped the development of the internet and related innovations. One such turning point was the Telecommunications Act of 1996, which dismantled long-standing monopolies in the telephone and cable industries. This move opened the door for increased competition and allowed new service providers to enter the market. As a result, internet access expanded rapidly, and high-speed connections became more affordable and widely available.

These shifts also gave rise to net neutrality debates, which centered on whether internet service providers should treat all online data equally or be allowed to prioritize certain services. This debate highlighted the ongoing tension between regulation and innovation, illustrating how legislation can simultaneously enable and constrain technological growth. It also underscored how policy decisions often reflect broader political, economic, and social dynamics.

Over time, the evolution of technology policy in the U.S. has played a critical role in shaping not only innovation but also the framework of modern governance. Understanding this historical context helps explain current policy directions and sheds light on the challenges and trends we see today.

The Telecommunications Act (1996)

In the 1990s, the United States underwent a dramatic shift in how technology and communications were regulated. The Telecommunications Act of 1996 marked a transformative era for the industry, becoming the first major overhaul of telecommunications law since 1934. This landmark legislation aimed to deregulate the sector by dismantling entrenched monopolies in telephone and cable services, opening both local and long-distance markets to competition.

The goal was clear: remove restrictions, invite new players, and drive innovation and affordability. The Act paved the way for a flood of Internet Service Providers (ISPs) like AOL, EarthLink, and thousands of smaller firms to rapidly expand, offering dial-up—and eventually broadband—access to millions of Americans. This surge in competition forced established providers to lower prices and improve services, accelerating internet adoption across the country.

The ripple effects were profound. With more households connected at faster speeds, the groundwork was laid for the explosion of web-based services, e-commerce, digital media, and the rise of tech giants that capitalized on the growing digital infrastructure (Internet: An Overview of Key Technology Policy Issues Affecting Its Use and Growth, n.d.).

At the same time, new debates emerged—particularly around net neutrality, which questioned whether ISPs should be allowed to prioritize certain types of data over others. This highlighted the double-edged nature of deregulation: while it fostered innovation and access, it also opened the door to power imbalances and new regulatory challenges.

Overall, the Telecommunications Act of 1996 brought both benefits and complications. It democratized access to the internet and stimulated a new digital economy—but it also introduced complexities that continue to shape tech policy today.

Telecommunications Act

Pros:

1. **Increased Competition:** The promotion of competition by reducing regulatory barriers led to more service options and better pricing for consumers. Eventually, it enabled massive access to the new services.
2. **Technological Innovation:** By encouraging competition, the Act spurred innovation, enabling advancements in broadband, mobile communications, and the expansion of Internet networks.
3. **Deregulation of Media Ownership:** The legislation relaxed restrictions on media ownership, allowing for consolidation and the growth of large media companies, potentially enhancing content diversity and accessibility.

Cons:

1. **Consolidation of Market Power:** While competition was encouraged, the Act also led to significant mergers and acquisitions, resulting in the dominance of a few large corporations that could stifle smaller competitors and limit consumer choices. AT&T, Verizon, Comcast, Time Warner, and Sprint, among others, leveraged the opportunities created by the Telecommunications Act to consolidate their positions in the telecommunications and media markets.
2. **Neglect of Rural Areas:** The market-oriented perspective focused on urban areas, often leaving rural and underserved regions with inadequate access to high-quality telecommunications services, exacerbating the digital gap.
3. **Concerns Over Content Regulation:** The Act raised fears regarding content regulation and free speech, as the deregulation

of media ownership could lead to biased information dissemination and a lack of diverse viewpoints.

The Digital Millennium Copyright Act (1998)

Enacted in 1998, the Digital Millennium Copyright Act (DMCA) was designed to adapt copyright law to the digital age, addressing challenges posed by the Internet's rapid growth. A key focus of the DMCA was to strike a balance between protecting intellectual property rights and promoting online innovation. It sought to encourage creativity while safeguarding the rights of content creators.

However, the implementation of the DMCA has not been without controversy. Critics argue that some provisions of the Act, such as those related to anti-circumvention measures, can stifle innovation and limit legitimate uses of digital content.

Net Neutrality Legislation

The Federal Communications Commission (FCC) introduced regulations specifically enforcing net neutrality. These regulations were designed to ensure that Internet service providers (ISPs) treat all data transmitted over the Internet equally, without favoritism or discrimination (Hayes, 2025). Here's a breakdown of what this means:

1. **Equal Treatment of Data:** ISPs are required to handle all Internet traffic in an impartial manner. This means they cannot block or slow down access to specific websites, applications, or services based on their content or the identity of the provider.

2. **Prohibiting Blocking:** ISPs are not allowed to completely block access to any legal content or sites. For example, an ISP cannot prevent users from accessing a competitor's streaming service or

advocating for a particular political viewpoint by blocking access to those platforms.

3. **Prohibiting Throttling:** Throttling refers to the intentional slowing down of Internet service. Under these regulations, ISPs cannot intentionally reduce the speed of specific content or services. For instance, if a user streams videos from a competing platform, the ISP cannot slow down that service's connection while keeping its own preferred service running at high speed.
4. **Disclosure Requirements:** The regulations also require ISPs to transparently disclose their network management practices, fees, and other aspects of their services to consumers, promoting informed decision-making.

The goal of these regulations was to prevent ISPs from engaging in unfair practices that could harm competition, limit consumer choice, and stifle innovation in online services and content. By ensuring a level playing field for all Internet traffic, the FCC aimed to protect the openness of the Internet and support a vibrant digital ecosystem.

Trends in Fintech Regulations

The 2008 financial crisis exposed deep vulnerabilities in the global financial system, prompting regulators to reevaluate long-standing policies. In the years that followed, a clear shift emerged toward tighter oversight of financial technology—an effort to balance stability and innovation as new tools transformed how consumers interacted with money.

This recalibration extended beyond financial systems to data privacy, particularly under the Trump administration's second term. Federal enforcement saw a sharp rollback: in June 2025, guidelines that had required explicit user consent before sharing or selling personal data were rescinded. Many consumer protections previously advanced by the Federal Trade Commission (FTC) were dismantled, reducing corporate obligations to disclose how user data was collected, stored, or used. As a result, major tech firms expanded their data aggregation and targeted advertising practices. One leading social media platform, for example, updated its default privacy settings to broader data sharing—triggering a surge in personalized ads. Critics, including privacy advocates, warned that users were losing meaningful control over their personal information, citing incidents where geolocation data was sold without notification, raising alarm about surveillance and misuse. Industry supporters, however, defended the changes as necessary for innovation, pointing to smoother onboarding and more tailored experiences in both fintech and health-tech apps (Larrison, 2025).

Simultaneously, the growth of fintech was further fueled by the expansion of regulatory sandboxes—controlled environments where startups could test financial products under relaxed legal conditions. These sandboxes enabled early-stage companies to experiment with peer-to-peer payment platforms, AI-driven fraud detection, and digital lending services without the full burden of compliance upfront. A standout example was ZapPay, a startup that piloted instant international transfers within a sandbox before scaling operations nationwide. Some sandboxes included regular oversight and consumer disclosure mandates, while others relied heavily on self-reporting. Supporters saw these models as vital for lowering entry barriers and encouraging rapid innovation. But critics pointed to gaps in liability protections and consumer safeguards, warning that without robust oversight, users could face unclear recourse in the event of errors or breaches (Larrison, 2025; Manfredi, 2025).

Ultimately, the rise of fintech presented both a challenge and an opportunity for regulators. As technologies like mobile banking, digital currencies, and smart contracts emerged, policy frameworks had to evolve in real time—adapting not just to support innovation, but to mitigate growing concerns around fraud, cybersecurity, and consumer protection.

Geographical Fintech Policies and Frameworks

Global

- **Financial Stability Board (FSB) Guidelines:** The FSB has released guidelines and recommendations for regulating fintech and ensuring that financial stability is maintained in light of new technologies and services.
- **Basel Committee on Banking Supervision (BCBS) Guidelines:** The BCBS has published principles for the effective management and supervision of fintech risks, urging banks to adapt their risk management frameworks to account for the unique challenges posed by fintech innovations.
- **Regulatory Sandboxes:** Many countries, including the UK, Singapore, and Australia, have established regulatory sandboxes that allow fintech start-ups to test their products in a controlled environment while ensuring compliance with regulatory standards.

European Union

- **Payment Services Directive 2 (PSD2) (2018):** This directive aimed to increase competition, innovation, and security in the payment services sector by allowing third-party providers to access bank data with customer consent, thereby enabling new fintech services.

- **General Data Protection Regulation (GDPR) (2018):** Though primarily focused on data protection and privacy, GDPR has significant implications for fintech companies in how they handle personal data, requiring compliance to ensure consumer protection.
- **Capital Markets Union (CMU):** An ongoing initiative to enhance access to finance for companies, particularly start-ups and fintechs, by creating a more integrated capital market within the EU.

United States

- **Dodd-Frank Wall Street Reform and Consumer Protection Act (2010):** This comprehensive legislation aimed to reduce risks in the financial system and included provisions for increased transparency and oversight of financial institutions.
  - ▹ The Act established the Consumer Financial Protection Bureau (CFPB) to oversee consumer protection in financial products and services, impacting fintech companies offering loans and payment services.
- **Banking Innovation Act (2019):** This legislation aimed to facilitate fintech development by encouraging banks to adopt new technologies and fostering collaboration between traditional banks and fintech start-ups.
- **Innovations in Technology (Office of the Comptroller of the Currency - OCC):** The OCC has issued frameworks and guidelines to explore how fintech companies can engage with the banking system, including the possibility of receiving special-purpose national bank charters.

These regulations and initiatives reflect a global trend toward enhancing oversight, promoting innovation, and ensuring consumer protection in the growing fintech sector since the 2008 financial crisis. The landscape continues to evolve as technology advances and new challenges emerge.

Compliance and Reporting Requirements

Post-2008, compliance and reporting requirements have become more stringent, emphasizing transparency and integrity in the fintech sector. The global financial crisis highlighted vulnerabilities and opaque practices within financial markets, prompting regulatory bodies to tighten norms to avert future catastrophes. Enhanced compliance standards require fintech companies to adhere to rigorous reporting obligations.

These regulatory measures serve two primary objectives: protecting consumers and preserving the credibility of the financial system. By promoting transparency, they help identify and mitigate risks associated with financial innovation before they escalate into systemic threats. For fintech firms, adhering to these standards is an opportunity to cement their reputation for reliability and trustworthiness. Compliance practices build a foundation for sustainable growth, offering firms a distinct competitive advantage in an increasingly crowded marketplace.

Another vital aspect of evolving fintech regulation is the push toward global harmonization of standards. With fintech companies operating across borders, cohesive international policies are crucial to facilitating cross-border operations and reducing associated risks. Harmonization efforts aim to create a level playing field where companies can operate consistently across different jurisdictions without facing conflicting regulations. Additionally, global alignment reduces regulatory arbitrage, wherein firms exploit loopholes by relocating to jurisdictions with more lenient standards. Through harmonized regulations, countries aim to support seamless fintech operations, ultimately driving innovation while safeguarding against systemic risks.

Scrutiny Over Cryptocurrencies

Cryptocurrencies have come under the lens of concerns about market volatility, potential fraud, and the implications for traditional financial systems. Governments worldwide are working to establish clear regulatory guidelines that enhance consumer trust and ensure market stability. This cryptocurrency regulation seeks to protect investors from fraudulent schemes while ensuring an environment where digital assets can be used safely.

Since the creation of cryptocurrencies, particularly Bitcoin, in 2009, various legislations and regulatory measures have emerged globally to address the unique challenges and risks associated with digital currencies.

Here's an overview of key legislation and regulatory developments regarding cryptocurrencies from their inception until 2018:

2008-2013: Early Developments

- **Bitcoin Whitepaper (2008):** Satoshi Nakamoto published the *Bitcoin Whitepaper* outlining the concept of a decentralized digital currency, laying the foundation for future cryptocurrencies but without formal regulatory oversight.
- **FinCEN Guidance (2013):** The Financial Crimes Enforcement Network (FinCEN) issued guidance stating that cryptocurrency exchanges and administrators must comply with anti-money laundering (AML) regulations, categorizing them as money transmitters.

2014: Regulatory Frameworks Begin

- **IRS Guidance (2014):** The Internal Revenue Service (IRS) announced that cryptocurrencies would be treated as property for tax purposes, meaning that capital gains tax would apply to cryptocurrency transactions.
- **New York BitLicense (2014):** The New York State Department of Financial Services (NYDFS) introduced BitLicense, a regulatory framework requiring businesses involved in virtual currency activities to obtain a license and comply with consumer protection and AML regulations.

2015-2016: Global Responses and Regulatory Initiatives

- **EU's 4th Anti-Money Laundering Directive (2015):** This directive included provisions for regulating cryptocurrencies, requiring member states to establish measures to combat money laundering and terrorist financing involving virtual currencies.

- **EU Court Ruling (2015):** In a landmark ruling, the European Court of Justice exempted Bitcoin transactions from value-added tax (VAT), recognizing their status as a currency.
- **Chinese Regulations (2016):** China began implementing stricter regulations on cryptocurrencies, banning initial coin offerings (ICOs) and shutting down domestic cryptocurrency exchanges.

2017: Rapid Growth and Increasing Scrutiny

1. **SEC Guidance (2017):** The U.S. Securities and Exchange Commission (SEC) issued guidance on ICOs, stating that many tokens sold in ICOs could be classified as securities and thereby subject to securities laws.
2. **CFTC Classification (2017):** The Commodity Futures Trading Commission (CFTC) clarified that Bitcoin and other cryptocurrencies are considered commodities, giving them regulatory oversight concerning derivatives trading and fraud.
3. **G20 Meeting (2017):** Finance ministers and central bank governors from G20 nations discussed the need for a coordinated approach to regulating cryptocurrencies, emphasizing oversight to mitigate risks to the financial system.

2018: Continued Regulatory Developments

- **EU's 5th Anti-Money Laundering Directive (2018):** This directive expanded the regulatory framework for cryptocurrencies to include stricter rules for cryptocurrency exchanges and wallet providers, aiming to enhance transparency and combat AML and terrorist financing.
- **Global FSB Report (2018):** The FSB published a report emphasizing the need for global regulatory coordination concerning

cryptocurrencies, recognizing their potential impact on financial stability.

With their rapid growth and widespread adoption, cryptocurrencies challenge conventional economic models. Regulatory clarity provides reassurance to consumers and businesses, encouraging further investment and innovation within the cryptocurrency sphere. The evolution of regulations surrounding cryptocurrencies underscores the importance of maintaining a flexible yet protective stance towards new financial technologies (Schwarcz, 2024).

Shifts in Policy Under Trump's First Term

During Donald Trump's first term (2017-2021), several changes in policy emerged, each shaping the landscape of technology governance in profound ways. **One of the most notable trends was the administration's emphasis on deregulation.** Trump's push to reduce regulatory burdens across various sectors, including technology, stemmed from a belief that less governmental oversight promotes growth and innovation.

By rolling back regulations, the administration aimed to stimulate the tech industry. Yet, this approach raised critical challenges regarding accountability and consumer protection. The strategic decision-making behind these policies revealed deeper ideological debates about the role of government in regulating industries, especially in the rapidly evolving tech landscape.

Shifts in Technology Policies

- **American Technology Council (2017):** Trump established the American Technology Council to modernize federal IT, improve cybersecurity, and promote collaboration between the federal government and private tech companies.

- **Industry Engagement:** The administration emphasized partnerships with leading technology companies to enhance government efficiency, with a focus on cloud computing, data management, and AI advancements.

Shifts in Fintech

- **OCC Initiatives:** The OCC under Trump encouraged fintech innovation through initiatives such as the "fintech charter," which allowed certain fintech companies to operate as national banks. This aimed to simplify federal regulations and encourage competition in the financial sector.
- **Encouraging Financial Innovation:** The administration supported the growth of financial technologies by advocating for deregulation and creating a more favorable environment for fintech firms to operate. This included efforts to facilitate partnerships between traditional banks and fintech companies.
- **CFPB Oversight:** Although the CFPB's focus under the Trump administration shifted toward reducing regulatory burdens, it continued to play a role in overseeing fintech companies and ensuring consumer protections in financial products and services.

These policies significantly contributed to deregulation efforts by prioritizing the modernization of federal IT systems and enhancing cybersecurity measures. By creating a dedicated council aimed at bridging the gap between the federal government and private technology firms, Trump's administration sought to bring innovation and increase efficiency within government operations. This initiative encouraged the implementation of cutting-edge technologies and streamlined processes, ultimately reducing bureaucratic red tape and allowing for more flexible regulatory frameworks that could adapt to the rapidly evolving technological landscape.

Furthermore, the administration's emphasis on industry engagement with leading technology companies played a crucial role in deregulation. The government-supported partnerships with private companies focused on cloud computing, data management, and advancements in AI. This collaborative approach resulted in the adoption of more efficient technological solutions but also led to the relaxation of certain regulations that previously hindered private sector involvement in government projects.

Shifts in Cybersecurity

- **Executive Order on Cybersecurity (2017):** Executive Order 13800, *Strengthening the Cybersecurity of Federal Networks and Critical Infrastructure,* was signed to improve critical infrastructure cybersecurity and promote private sector engagement in cybersecurity initiatives. It called for a review of the cybersecurity policies of federal agencies and encouraged the adoption of best practices.
- **Cyber Strategy (2018):** The Trump administration released the National Cyber Strategy in an effort to enhance cybersecurity across government, defense, and private sectors. The strategy prioritized deterrence against cyber threats, strengthening critical infrastructure protections, and encouraging information sharing between the public and private sectors.
- **Cybersecurity and Infrastructure Security Agency (CISA):** Established under the Department of Homeland Security (DHS) in 2018, CISA focused on protecting the nation's critical

infrastructure from cyber threats and providing cybersecurity resources for state and local governments.

While such policies encouraged innovative tech solutions, often in response to security threats, they also sparked discussions about protectionism and market access. Understanding the national security tech policies introduced during this time is essential for grasping the contemporary U.S. political environment.

Shifts in Cryptocurrency

- **Regulatory Approach:** The administration generally took a cautious approach toward cryptocurrencies, focusing on preventing illegal activities such as money laundering and fraud while acknowledging the innovation potential. The SEC and CFTC increased scrutiny of ICOs and cryptocurrency markets.

This approach was meant to study the implications of cryptocurrencies and blockchain technology, seeking to understand the potential benefits and risks associated with digital currencies. The administration was looking forward to informing future regulatory decisions and crafting a comprehensive stance on cryptocurrencies. This proactive examination allowed for a more nuanced regulatory framework with a focus on supporting innovation rather than addressing concerns related to security and compliance.

Moreover, the administration's generally cautious regulatory approach toward cryptocurrencies facilitated a balanced environment that permitted the exploration of new financial technologies while addressing necessary legal safeguards. While there was a discussion on preventing illegal activities such as money laundering and fraud, the administration recognized the need for oversight without imposing overly stringent regulations that could stifle innovation. Increased scrutiny from agencies like the SEC and CFTC regarding ICOs indicated a willingness to regulate where necessary but also highlighted an openness to allow legitimate cryptocurrency enterprises to thrive.

Shifts in Trade Policies

During Donald Trump's presidency, several key policies and legislative actions were taken regarding technology trade and relations with China. Here's an overview of those actions:

- **Tariffs on Chinese Goods (2018-2019):** The Trump administration imposed tariffs on a wide range of Chinese imports, including technology products, under the rationale of addressing unfair trade practices and intellectual property theft. These tariffs aimed to protect U.S. companies and encourage domestic manufacturing.
- **Section 301 Investigations (2018):** In 2018, the U.S. Trade Representative (USTR) initiated Section 301 investigations into China's trade practices, particularly focusing on issues related to intellectual property theft and forced technology transfers. This investigation led to the imposition of tariffs on Chinese goods.
- **Entity List and Export Controls:** The administration placed several Chinese technology companies, including Huawei and ZTE, on the Entity List, restricting their access to U.S. technology and components. This move was part of broader national security concerns regarding potential spying and cybersecurity threats posed by these companies.

- **Clean Network Initiative (2020):** Launched in 2020, the Clean Network Initiative aimed to protect U.S. citizens' privacy and prevent Chinese companies from infiltrating sensitive information, promoting the use of trusted vendors and technologies in 5G networks and other critical telecommunications infrastructure.

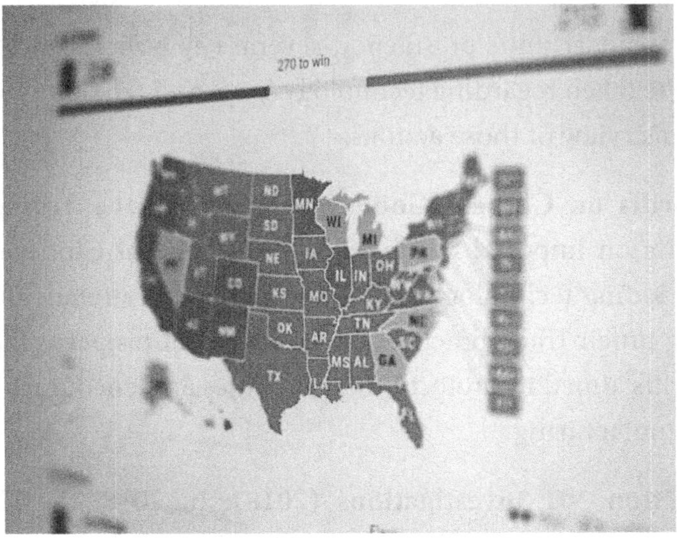

Shifts in Policy Under Biden Administration

Under the Biden administration (2021-2025), policies took a different direction, particularly in the realm of AI and fintech.

Shifts in Artificial Intelligence

- **Emphasis on Ethical AI:** The Biden administration took a new approach that prioritized the development of responsible and ethical AI practices, promoting fairness, accountability, and transparency in AI systems. Initiatives aimed to mitigate biases and protect civil rights in AI applications.

- **Investment in Research and Development:** Funding for AI research and innovation was increased to maintain the U.S. competitive edge in AI technology. This included collaborations between federal agencies, academic institutions, and the private sector.
- **National AI Strategy:** The administration worked on a comprehensive national AI strategy that included workforce development, education, and partnerships to ensure that the technology benefits all Americans while addressing potential societal impacts.

Shifts in Fintech Policy

- **Enhanced Regulatory Oversight:** To increase regulatory scrutiny in the fintech sector, the Biden administration introduced greater oversight, focusing on consumer protections and addressing risks associated with digital financial services, including cryptocurrencies.
- **Support for Financial Inclusion:** There was a strong emphasis on promoting financial inclusion through fintech innovations, particularly for underserved communities.
- **Sustainability and Innovation:** Efforts were made to recognize the role of fintech in combating climate change and promoting sustainable finance, encouraging innovations that align with environmental goals.

Overall, Biden's administration opted for a more proactive and regulated approach to AI and fintech, emphasizing ethical considerations, consumer protections, and inclusivity while encouraging innovation and collaboration across sectors. The promotion of transparency in AI algorithms is critical to ensure fair and accountable outcomes. This goes hand-in-hand with protecting privacy and data security in AI applications, as the potential risks associated with misuse demand rigorous oversight.

On an international scale, the Biden administration strengthened global partnerships by engaging in dialogues with international regulators and collaborating on shared standards for AI ethics. Addressing the global implications of AI and fintech, the administration sought to navigate challenges related to cross-border data flow and technological access disparities among countries.

Impact of Deregulation on Innovation

The story of telecommunications deregulation provides a clear example of how reducing regulatory barriers can fuel industry expansion. The Telecommunications Act of 1996 is illustrative in this regard. While restrictions allowed more players in the market, they also led to market consolidation, where a few large companies absorbed many smaller ones. As these dominant players grew, concerns about monopolistic practices emerged. This tension between encouraging innovation and preventing monopoly formation remains relevant today, particularly in discussions surrounding major tech giants.

Similarly, the financial sector has witnessed dramatic changes due to deregulation, especially with the rise of fintech and non-traditional banking services. Deregulation provided an entryway for start-ups and agile innovators, promoting diverse financial products and services that cater to the evolving needs of consumers. However, the absence of stringent regulations also brought systemic risks. The 2008 financial crisis exposed these vulnerabilities starkly. With less oversight, risky lending practices became prevalent, ultimately contributing to the global financial meltdown.

Certain economic sectors have significantly benefited from recent technological innovations. For instance, in agriculture, deregulation has significantly influenced technological advancements, particularly in biotechnology. For example, the drive to remove the barriers allowed for the accelerated development of genetically modified organisms (GMOs), which aimed to enhance crop yield and resistance. These innovations promised increased efficiency and food security. Yet, they also sparked intense debates regarding their safety, ethical implications, and environmental impact.

Conversely, other sectors of society pay the price. Labor markets, for instance, are profoundly affected by deregulation. On one hand, deregulation can lead to job creation, as seen in the tech sector's continuous growth through innovation. As new industries emerge, they offer employment opportunities and demand novel skill sets. This evolution necessitates workforce adaptation, as traditional roles may become obsolete or transform significantly. However, deregulation can also result in job displacement, particularly in fields that fail to adapt quickly enough to technological change.

Reflecting on this landscape, readers are invited to consider: What mix of freedom and protection should define tomorrow's tech policies? How do you weigh the benefits of rapid progress against the need for accountability? These are questions with no easy answers, but understanding the history and ongoing dialogue is a step toward meaningful solutions.

Bringing It All Together

- Tech legislation has shifted between deregulation and increased oversight, balancing innovation with individual rights.

- The early 2000s saw minimal regulation, fueling rapid internet growth. Rising concerns about privacy and monopolies later triggered stronger consumer protections.
- Trump's first term focused on deregulation and protecting U.S. industries. Biden's term emphasized ethics, consumer protection, and financial inclusion.
- With Trump's return in 2025, tech policy pivoted again:
  - ▹ Revoked Biden's AI executive order.
  - ▹ Prioritized U.S. AI dominance through deregulation.
  - ▹ Launched a federal investigation into content moderation.
  - ▹ Announced the $500B "Stargate Project" to boost AI infrastructure.
  - ▹ Formed a task force to revise cryptocurrency rules.
- These moves aim to reduce barriers and boost private sector innovation, but raise concerns about oversight and fairness.
- Global Legislation:
  - ▹ Europe enforces strict rules (e.g., GDPR, AI Act).
  - ▹ Asia varies—China enforces control; Singapore and Japan offer flexible models.
  - ▹ Latin America takes a pragmatic, evolving approach.
- As tech grows more global, regulatory harmonization becomes a major challenge.
- Key questions remain: How much freedom should tech firms have? How do we balance innovation with accountability?

Chapter 3

Trump's Executive Orders and Federal Policies

"Make America Great Again," the MAGA movement, was Trump's political campaign strategy in 2016 and was revitalized when he launched his candidacy for the presidency in 2024. It encapsulated his vision for reinforcing the U.S. economy and restoring national pride, resonating deeply with voters who felt left behind by the rapid technological changes. This slogan underscored his commitment to prioritizing U.S. interests, particularly in the context of job creation and economic stability. A few days after taking office, Trump passed his first executive order to fulfill his campaign promises.

In his second term, Trump's approach includes a significant shift toward tech deregulation, wherein he argues that reducing government oversight would spur innovation and economic growth, allowing U.S. businesses to compete more effectively on the global stage. The White House stated about Executive Order 14179 in the following terms: "This Executive Order establishes the commitment of the United States to sustain and enhance America's dominance in AI to promote human flourishing, economic competitiveness, and national security" (The White House, 2025e, para. 8). This philosophy shapes his administration's policies, seeking to balance innovation with nationalistic economic goals, thus influencing the broader discourse on the role of technology in the future of the US.

Trump's executive orders and federal policies reveal a complex interface between governance and technological innovation. The era of the Trump administration marks a distinct moment when technology takes center stage in federal policymaking, shaping how governments interact with emerging technologies like cybersecurity, AI, and digital finance. With increasing cyber threats and the rapid evolution of AI and blockchain technologies, the need for clear directives has become more pronounced. The executive orders Trump has issued during this first period seek to address the opportunities and challenges posed by these advancements. They aim to reinforce national security while fostering an environment that encourages technological progress without compromising ethical standards or consumer privacy.

Trump's Initial Key Executive Orders and Policies

President Trump's second term marked a sharp pivot in federal tech governance. Departing from the Obama-Biden era's focus on compliance-heavy modernization, his administration embraced a streamlined, pro-innovation agenda centered on targeted investment, decentralization, and a reduction in federal oversight. The strategy aimed to bolster American technological dominance while aligning with populist concerns over government overreach (Sayegh, 2025).

Within his first week in office, Trump issued a series of executive orders that set the tone for his digital policy agenda:

- **Executive Order 14158 *Implementing the President's "Department of Government Efficiency" Cost Efficiency Initiative*:** On January 20, 2025, an executive order renamed the United States Digital Service as the United States DOGE (Department of Government Efficiency) Service. This rebranding aimed to modernize federal technology and software to maximize governmental efficiency and productivity.
- **Executive Order 14178 *Strengthening American Leadership in Digital Financial Technology*:** On January 23, 2025, this executive order was issued to promote U.S. leadership in digital assets and financial technology.
- **Executive Order 14179 *Removing Barriers to American Leadership in Artificial Intelligence*:** Also on January 23, 2025, President Trump signed this executive order to enhance America's global AI dominance. This order revoked prior policies considered to hinder AI innovation and directed the development of an AI action plan to sustain and enhance U.S. leadership in AI.

Each executive order entailed a number of public policy shifts that will be explored and explained next.

Revocation of Previous Orders

Executive Order 14178 revokes Executive Order 14067, which aimed at ensuring responsible digital asset development, and cancels related Treasury documents. Signed under the Biden administration, Executive Order 14067 emphasized precautionary measures to address issues like illicit finance, financial stability, and consumer protection in the digital asset space.

Additionally, Executive Order 14178 calls for the immediate revocation of the Framework for International Engagement on Digital Assets, issued by the Treasury Department in July 2022. This framework was designed to facilitate U.S. engagement with global partners on the regulation, innovation, and enforcement related to digital assets, including combating illicit finance and enhancing cooperation with international organizations. By revoking these measures, President Trump is signaling a pivot toward a more domestically focused agenda that prioritizes deregulation, financial autonomy, and technological innovation over international collaboration and the exploration of central bank digital currencies (CBDCs).

Prohibition of Central Bank Digital Currencies

Executive Order 14178 enacts a clear prohibition on federal agencies from establishing, issuing, or promoting CBDCs within the US and internationally. This directive requires the immediate cessation of any ongoing plans or initiatives related to the development of CBDCs in the US.

This prohibition aligns with President Trump's campaign rhetoric, wherein he expressed strong opposition to the creation of CBDCs, arguing that it would enable the federal government to exert "absolute control" over individuals' finances. Concerns raised by Trump and other critics focus on the potential for increased government surveillance of Americans' financial transactions, which they believe could undermine personal privacy, threaten individual sovereignty, and destabilize the financial system (*President Trump Issues Executive Order*, 2025).

In contrast, various other nations are progressing with their own CBDC initiatives; for instance, the European Central Bank is set to unveil critical features of its digital euro, highlighting a divergence between U.S. policy and international trends in digital currency development.

Support for Digital Assets

Executive Order 14178 emphasizes the importance of the digital asset industry for innovation and economic growth, advocating for responsible development and use of digital assets and blockchain technology.

The order creates a President's Working Group on Digital Asset Markets, made up of the heads of several federal agencies, including the SEC AND CFTC, and led by the President's Special Advisor for AI and Crypto. This working group's primary responsibility is to generate regulatory and legislative recommendations in a timely manner. Within 30 days, agency heads are required to identify all existing regulations, guidance, and orders that impact the digital asset sector. Subsequent to this, agency heads must submit proposals for rescinding, modifying, or adopting specific regulations to the working group's chair within 60 days.

Within 180 days of the executive order, the working group must deliver comprehensive recommendations to President Trump, including suggestions for a federal regulatory framework for digital assets, particularly focusing on stablecoins, and the potential creation of a "national digital asset stockpile" consisting of cryptocurrencies seized through law enforcement actions. The executive order also allows for public hearings and invites insights from experts in digital assets and markets. Following this directive, SEC Commissioner Hester Peirce, heading the SEC's newly established Crypto Task Force, outlined key priorities aimed at improving the regulatory landscape for cryptocurrencies. She criticized the SEC's past handling of crypto regulation, emphasizing the need for industry collaboration and open communication with the agency to effectively shape the regulatory approach moving forward.

Other relevant aspects to consider from Executive Order 14178 are (The White House, 2025f):

- **Citizen and Business Protections:** It ensures individual and business access to public blockchain networks, allows lawful transactions without censorship, and protects their self-custody rights.
- **Sovereignty of the U.S. Dollar:** The order aims to uphold the dollar's sovereignty, supporting lawful dollar-backed stablecoin development globally.
- **Access to Banking Services:** It promotes fair access to banking for all law-abiding entities and individuals.
- **Regulatory Clarity:** The executive order calls for technology-neutral regulations that adapt to new technologies to support a vibrant digital economy.
- **Creation of a Working Group:** The order establishes the President's Working Group on Digital Asset Markets to assess and recommend regulatory frameworks for digital assets.

- **Regulatory Review Timeline:** Agencies involved must identify and review existing regulations impacting digital assets, with recommendations for changes due within specific timeframes.
- **Severability and Implementation:** Provisions ensure that if part of the order is invalidated, the rest remains effective, and it confirms adherence to existing laws and budgetary guidelines.

Overall, the executive order's objective is to boost innovation in digital assets while safeguarding U.S. economic interests and individual rights.

Support for Artificial Intelligence

Executive Order 14179 instructs the Assistant to the President for Science and Technology, along with other senior officials from the White House and heads of relevant agencies, to undertake several key actions.

Firstly, they are tasked with developing a comprehensive plan within 180 days to align with the new U.S. policy on AI as outlined in the executive order. This policy aims to "enhance America's global AI dominance" (The White House, 2025d, sect. 2) as the key to human flourishing, boosting economic competitiveness, and strengthening national security.

Secondly, they will evaluate all actions taken by various agencies under President Biden's AI Executive Order 14110 from 2023, with the goal of rescinding any measures that conflict with the objectives set forth in Trump's new order.

Lastly, the directive calls for the revision of two Office of Management and Budget (OMB) memoranda (M-24-10 and M-24-18), which pertain to the government's acquisition and implementation of AI technologies, to ensure they are in line with the administration's renewed focus on optimizing AI use across federal agencies.

Foundations of the Executive Orders

President Trump's executive orders are a decisive initiative aimed at reinstating and enhancing the U.S. position as a global leader in digital innovation. Executive Order 14179 revokes the previous Biden Administration's AI policy, which President Trump characterized as overly restrictive and detrimental to private sector innovation. The Biden AI Executive Order 14110 imposed burdensome regulations that hindered the development and deployment of AI technologies, potentially stifling American technological advancements and competitiveness. By eliminating these obstacles, the Trump administration seeks to foster an environment where innovation can thrive unencumbered by excessive government intervention.

Furthermore, Trump's recent executive orders emphasize the importance of AI in promoting economic growth and national security. They reaffirm the commitment to develop AI systems free from ideological biases and social agendas, aiming to ensure that American technological advancements benefit all citizens. Executive Order 14179 outlines a plan for establishing an AI Action Plan to sustain and strengthen U.S. leadership in this critical area. It also mandates revisions to existing government policies related to AI acquisition and governance to eliminate any remnants of the prior administration's restrictive measures. By building on prior efforts and achievements in AI development, President Trump's directive strives to create a robust framework that prioritizes innovation and prepares the US to face future challenges in the global AI arena.

Main Concerns and Complaints About Trump's Executive Orders

Trump's executive orders to roll back existing AI protections have raised significant concerns regarding the potential negative impact on civil rights and public safety. By swiftly repealing Biden's initiatives aimed at ensuring the safe and fair use of AI, President Trump has prioritized rapid AI development over necessary safeguards that prevent bias and discrimination. Without these protections, there is a growing fear that AI tools will be used in critical decision-making processes—such as hiring, lending, and law enforcement—without sufficient oversight, leading to unfair outcomes. For example, automated systems may unjustly deny job opportunities or financial assistance based on flawed algorithmic assessments, exacerbating inequalities and resulting in real harm to individuals and communities.

The implications of these changes are particularly concerning for specific economic groups, such as workers in low-wage jobs or those seeking government assistance. The rollback of AI regulations signals a new direction in favor of large technology companies, which often have the resources to develop and deploy AI solutions without the same level of scrutiny as smaller firms. This dynamic can lead to reliance on untested and potentially harmful technologies, further marginalizing individuals who interact with increasingly automated government systems.

Furthermore, prominent figures in Silicon Valley have shown strong support for the Trump administration's approach, suggesting a convergence of interests where powerful tech leaders promote rapid AI deployment even at the expense of accountability and fairness. This marriage of economic influence and political power raises questions about the democratic process and the direction of technological advancement in the US (Faure, 2024).

Impact on Technology Companies

Establishment of The Department of Government Efficiency

President Trump's Executive Order 14158 establishing the DOGE and his focus on modernizing federal technology and software will have profound implications for technology companies operating in the US. The incorporation of the U.S. Digital Service into DOGE provides a dedicated team of technical experts aimed at enhancing government operations through innovative technologies. This could lead to improved efficiency in government processes, opening up opportunities for tech companies to partner with the federal government in developing new solutions and technologies.

However, the executive order also signals adjustments in the regulatory landscape for technology firms. The rescission of Biden's AI executive order and the emphasis on deregulation may alleviate compliance burdens for tech companies, creating a more advantageous environment for innovation and growth. Despite these potential benefits, firms may face challenges navigating the administration's commitment to eliminating ideological bias from tech platforms. This could compel companies to alter their moderation practices or algorithms to align with the administration's directives, potentially undermining user trust and satisfaction.

The Shift in Digital Asset Markets

The establishment of the Working Group on Digital Asset Markets represents a strategic pivot toward supporting the digital asset economy, distancing U.S. policy from concerns regarding fraud and money laundering. Technology companies focused on digital assets may find new avenues for growth, backed by an administration that prioritizes competitive advantages in this emerging sector. However, the effectiveness of these policies will depend significantly on how well the administration balances its goals of innovation with the need to protect consumer rights and national security. As technology companies adapt to these changes, maintaining open lines of communication with the federal government will be essential to navigate the evolving regulatory framework while maximizing opportunities for growth.

The regulatory impact on innovation represents a crucial consideration for technology companies adapting to the current situation. Trump's approach to deregulation seeks to reduce barriers to technological advancement. However, this deregulation must be carefully balanced with the need for oversight to ensure that rapid technological progress does not come at the expense of security or ethical concerns. Companies must find ways to innovate within the confines of any remaining regulations, adapting their strategies to thrive in an environment that simultaneously demands compliance and encourages creativity (Dunn, 2025a).

Possible Impact on Market Stakeholders

Another essential aspect to consider is the market's reaction to these changes and how companies, especially start-ups, adapt. The upcoming policies have led to varied investment trends as stakeholders reassess their strategies in light of new regulatory conditions. Start-ups may face unique challenges, as the regulatory framework governing their operations can change rapidly under executive orders. This environment demands agility from start-ups, requiring them to continuously adapt their business models to remain compliant while fostering innovation (Dunn, 2025a).

For businesses operating in this constantly evolving environment, there's also the element of strategic planning for long-term success. Companies must proactively manage their cybersecurity measures, adopting industry best practices such as multifactor authentication, encryption, and regular system assessments. By doing so, they not only align with federal expectations but also build resilience against potential threats, securing their digital assets and maintaining their competitive edge (Pearl, 2025).

Moreover, the ongoing impact of executive orders on market dynamics shows the importance of tech industries engaging in informed lobbying efforts. By staying actively involved in policy development, companies can help shape the regulatory landscape to better align with their interests and promote sustainable innovation. Engaging with policymakers provides a platform for tech companies to voice their concerns and advocate for policies that support their growth and strategic goals.

Corporate Compliance and Security Standards

As technology becomes more integral to national security, compliance with federal cybersecurity regulations has become paramount for tech companies. These regulations often require substantial alignment efforts, compelling corporations to invest heavily in their security infrastructure. The executive orders necessitate minimum standards such as multifactor authentication, endpoint detection and response tools, and data encryption (Pearl, 2025). While these measures enhance national security, they can also lead to increased operational costs for companies striving to meet compliance requirements.

The relationship between technology firms and government agencies is another area where Trump's executive orders leave a significant mark. These partnerships, aimed at bolstering national security, present both opportunities and challenges. On the one hand, collaborating with government entities can provide companies with access to real-time threat intelligence and resources that are typically unavailable to private firms. For instance, enhanced information sharing can help tech companies strengthen their defenses against emerging cyber threats, minimizing potential risks (Pearl, 2025). On the other hand, navigating public-private collaborations requires careful handling of sensitive information and adherence to stringent federal guidelines, which can be challenging and resource-intensive.

Changes in Data Privacy Regulations

Trump's legislation initiating significant changes to data privacy regulations marks a pivotal moment for consumer privacy rights in the US. This approach has stirred considerable debate among consumers, businesses, and advocacy groups. Under this administration, data privacy regulations reflect a strategic choice to prioritize national security concerns over individualized protections, which translates into a regulatory environment where the accessibility of personal data for both domestic and foreign entities is broader than before. Consumers now face a scenario where the government has more resources to scrutinize and oversee their privacy. This might make them feel uncertain about the safety of their personal information.

The Department of Justice (DOJ) introduced rules to restrict U.S. persons from engaging in data transactions involving sensitive personal information with "countries of concern," including adversarial nations. These rules, which are set to take effect in April 2025, emphasize national security by limiting access to sensitive data by foreign entities. However, the existence of the "regulatory freeze" raises uncertainty regarding their implementation, as it allows for potential delays and further stakeholder engagement regarding the rules (Jacobson et al., 2025).

Consequences for Consumers and Companies

The impact of these changes is twofold. For consumers, there is a growing apprehension regarding data management, particularly as the government's prioritization of security can lead to the monitoring and processing of personal data without sufficient protection. This anxiety is compounded by the reality that privacy rights, which many have assumed to be guaranteed, are increasingly subject to interpretation and regulatory changes that favor domestic interests and security measures over individual data protection. As consumers grapple with these new challenges, the disconnect between their expectations for privacy and the prevailing regulatory framework can diminish trust in both government and private sector entities.

For businesses, the new regulatory framework demands adaptations in data handling practices. Companies must now navigate a complex web of federal and state regulations, with states like California implementing stringent laws that can clash with federal directives. This intricate legal environment creates added pressure on businesses to comply with evolving data protection standards. Noncompliance can lead to severe legal consequences, positioning consumer trust as a critical asset that companies must actively cultivate by enhancing their data security measures. Furthermore, the new focus on cybersecurity compliance pushes companies to invest in robust systems that align with the dynamic regulatory framework while meeting consumer demands for accountability and transparency.

This period of enactment of the law also saw attention directed toward rules concerning recurring subscriptions and negative option programs, which aim to protect consumers from deceptive practices. Concerns raised regarding this rule included vague definitions of "unfair" actions and the need for better clarity on consumer rights.

In this changing context, businesses must balance compliance with conflicting regulations at both federal and state levels. As advocacy for consumer rights intensifies, the administration's approach will likely face scrutiny, emphasizing the ongoing need for clear, transparent policies that protect consumers while considering national security interests. The future of data privacy regulation in the US appears to be marked by contention and complexity as stakeholders from various sectors advocate for balanced approaches to safeguard individual rights in the digital age.

Consumer advocacy groups have emerged as key players in this evolving landscape, demanding clarity, transparency, and stronger protections against negligible oversight. Their calls for reform show a societal push toward balancing national security interests against the fundamental rights of individuals to safeguard their personal information. This ongoing dialogue emphasizes the need for policymakers to reconcile these competing priorities by crafting legislative frameworks that respect both security imperatives and consumer privacy rights.

Implications for Private Data

Under President Trump's administration, significant changes in federal data privacy regulations have been initiated, particularly with the executive memorandum titled *Regulatory Freeze Pending Review*, issued on January 20, 2025. This order halted the implementation of new regulations and mandated the withdrawal of any pending rules that had not yet been published as final in the Federal Register. Notably, this included amendments to the Children's Online Privacy Protection Rule (COPPA), which had been finalized by the Federal Trade Commission (FTC) shortly before the freeze. While the FTC considered these amendments to enhance protections for children's personal information, the order effectively stalled their implementation until a new head appointed by the president reviews and approves them.

The COPPA Rule amendments proposed several changes aimed at strengthening data privacy protections for children. These included requirements for operators to disclose to parents which third parties could access children's personal data and prohibiting the indefinite retention of such information. However, criticisms from FTC Chairman Andrew Ferguson pointed out significant flaws, such as unclear definitions and overly restrictive measures that could inadvertently harm users by risking excessive data deletion (Dunn, 2025b). For instance, the proposed prohibition on indefinite data retention could lead to the loss of valuable personal memories, as the term "indefinite" lacked precise interpretation, spreading confusion among stakeholders.

Moreover, Trump's executive memorandum has implications for specific sectors, particularly in health care, as it emphasizes restrictions on the transfer of personal health information (PHI) across borders. This increased scrutiny of foreign data access and the potential creation of conflicts with existing regulations such as HIPAA signify a transition toward greater national control over sensitive data types (InCountry Staff, 2025). Focusing on national security in data transfers can lead to strict controls that make it harder for businesses to comply with international operations. This may push companies to rethink their data residency and retention strategies while dealing with various regulatory environments.

Bringing It All Together

- At the start of his second term, President Trump signed multiple executive orders signaling a major shift in economic and political strategy.
- These policies have direct implications for both market stakeholders and everyday citizens.

- Critics warn that the loosely regulated framework introduced by these changes may:
  - ▷ Widen existing inequalities,
  - ▷ Disproportionately impact vulnerable populations,
  - ▷ And weaken long-standing consumer protections.
- The growing intersection of AI and political systems raises urgent questions about:
  - ▷ Democratic integrity in the face of rapid digital change,
  - ▷ The influence of technology-driven communication on public discourse,
  - ▷ And the need for ethical safeguards to preserve core democratic values.

Chapter 4

Artificial Intelligence Uses in Communication, Information, and Political Campaigns

It isn't coincidental that the world's greatest powers are increasingly concerned about the rapid advancements in AI and emerging technologies in information and communication. AI is reshaping the landscape of communication, information dissemination, and political campaigns. The way campaigns connect with voters, analyze preferences, and strategize outreach is creating an unprecedented scenario. AI isn't just a tool—it's becoming an integral part of the political process. From micro-targeting specific voter groups to crafting tailored messages based on sophisticated data analytics, AI provides unprecedented insights into the electorate's minds and behaviors. These technological advancements invite us to explore how they alter traditional campaigning methods and reflect on their implications for people's political engagement and democracy.

As countries try to improve their status in a competitive world, they are investing a lot in research, development, and infrastructure for AI and new technologies. The US and China are at the forefront, aiming to boost their economies and strengthen national security. This focus on innovation shows that these technologies will play a big role in the future, leading countries to make smart investments to benefit from AI and information systems.

Recent changes in AI investment and rules in the US mark a new phase for the industry. Increasing competition between the US and China is increasingly influenced by private companies in technology development, mixing national security with business interests. Overall, these developments suggest a significant change in the connection between technology, economy, and global strategy.

Let's explore the changes that President Trump has driven in his second term in office as well as the concerns these changes raise. What can be the potential outcomes of the new policies?

Artificial Intelligence New Policies After Trump's Executive Order

President Trump's administration has introduced significant changes to the AI landscape. Executive Order 14148 ended some programs from the Biden administration, including a 2023 order for protecting American citizens in AI development. This decision shows Trump's push for less regulation, claiming that past rules slowed tech progress. At the same time, the government announced the Stargate partnership, a $500 billion plan to improve U.S. AI infrastructure, which includes building data centers and clean energy facilities to support AI growth.

Further solidifying Trump's AI policy direction, Executive Order 14179, issued on January 23, emphasizes sustaining America's global dominance in AI and directs policy leaders to devise an action plan within 180 days. This order seeks to review and potentially retract any Biden-era policies that conflict with the new vision for AI.

Collectively, these actions signify a robust push toward innovation and a shift away from regulatory frameworks, redefining how federal agencies will manage and engage with AI technologies in the evolving economic landscape.

These measures have sparked debates in international policy and market arenas, positioning the US in a controversial stance against global regulatory frameworks. By challenging the economic and technological policies of other nations, particularly China, the US is amplifying tensions in the global race for technological supremacy.

A Tech-War? Trump's Policies Against Chinese Companies

Trump's policy toward China's tech companies reflects a strategic approach to curb their influence and competition in the global market. By implementing restrictions and promoting domestic investments in technology, the administration seeks to safeguard national security while bolstering American innovation against rival Chinese firms.

The Future of TikTok in the United States

President Trump's policy toward TikTok has been marked by efforts to manage its presence in the US while addressing national security concerns related to its Chinese ownership by ByteDance. In January 2025, Trump signed Executive Order 13942, which granted TikTok a 75-day extension to comply with a law mandating its sale or face a ban.

Trump temporarily halted enforcement of the ban while exploring avenues for a potential joint venture. His suggestions, including a 50-50 partnership between ByteDance and U.S. interests, indicate a desire for a compromise that allows TikTok to continue operating. However, this approach opposes a Supreme Court ruling aimed at protecting user data from Chinese control.

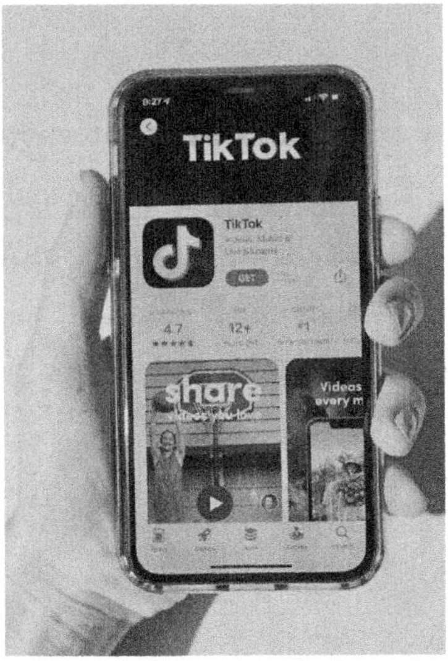

China may not benefit significantly from this arrangement, as it reinforces the perception of its companies as security threats, which may encourage ongoing scrutiny and opposition. Meanwhile, U.S. companies like Oracle and major influencers like Elon Musk and Kevin O'Leary have shown interest in acquiring TikTok, positioning themselves as potential winners in this scenario.

Additionally, if TikTok's future remains uncertain, platforms like Instagram Reels and YouTube Shorts could capitalize on the new context, redirecting advertising dollars away from TikTok. This interaction between companies highlights the interplay of national security, corporate interests, and consumer engagement, and the role played by states.

DeepSeek's Rise: Trump's Strategic Response to a New Artificial Intelligence Contender

In the first months of 2025, the emergence of the Chinese tech company DeepSeek led to an immediate response from President Donald Trump. On January 19, China unveiled an AI investment fund as a direct response to the tightened U.S. export controls on semiconductor chips.

Just two days later, Trump announced the Stargate Project, a bold initiative set to invest an unprecedented $500 billion into the development of U.S. AI infrastructure (Wilkinson, 2025). This ambitious project is backed by notable technology firms such as OpenAI and Oracle, alongside significant financial support from the Japanese bank SoftBank and the Emirati sovereign wealth fund MGX. The subsequent week saw DeepSeek unveil its low-cost, open-source AI model with advanced reasoning capabilities, which created market turmoil when it became evident that it posed a significant challenge to the US-based AI company, o3. As a result, NVIDIA, a leading chip manufacturer, witnessed a staggering $600 billion drop in market value when the markets opened on January 27 (Subin, 2025).

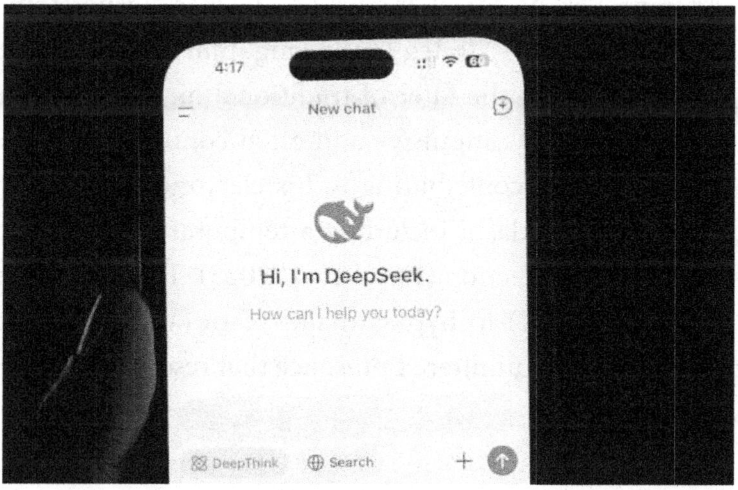

DeepSeek's innovations have challenged the previously held beliefs about the exclusivity in powerful AI development, raising questions about the effectiveness of stringent U.S. chip export controls. Beyond merely intensifying the competitive landscape between the US and China, these developments signify a substantive shift in U.S. tech governance toward a heavy concentration of power within the private sector regarding technology development and agenda-setting.

The scale of Stargate hints at an increasing intersection of national security interests, infrastructure, and Silicon Valley aspirations. OpenAI's economic framework for American reindustrialization—apparently a vital influence on Stargate—has emerged as an exceptionally powerful force, advocating for classified facilities to assess model security and necessitating new energy resources to effectively cool data centers. Trump's approach signals an emphasis on AI and technology, revealing an environment where national interests and innovative capacities intertwine.

Artificial Intelligence and Social Media as the New Way of Political Communication

Donald Trump's use of social media has been a cornerstone of his political communication strategy, allowing him to connect directly with supporters without the filter of traditional media. His reliance on Twitter, characterized by impulsive and often controversial messaging, has been credited with contributing to his electoral success while also causing significant backlash, including a temporary ban following the January 6 Capitol insurrection (Rasul et al., 2025). Trump has effectively utilized Twitter (now X) to bypass mainstream outlets and target his base, promoting a bold, unfiltered presence that resonated strongly with many Americans.

However, his tweets have also been linked to incivility and radicalization, creating a toxic discourse that critics argue undermines democratic norms. Researchers have noted that Trump's communication often employs moral disengagement mechanisms, justifying behaviors that would typically be deemed unacceptable (Torossian, 2025). Trump's focus has moved to his own platform, Truth Social, where he continues to engage with his loyal following. This evolution signifies a commitment to reinforcing direct communication while utilizing advanced technologies to shape public perception.

As he navigates a fragmented media landscape, Trump's strategy will likely incorporate innovative digital tools, enabling him to maintain a commanding narrative despite the challenges of political opposition and media scrutiny. Thus, Trump's communication approach is an example of a new paradigm in political discourse, and it raises questions about how this impacts civility, democratic engagement, and the evolving role of technology in politics.

Truth Social: Trump's Own Social Media

Truth Social is a social media service launched in 2022 by Trump Media & Technology Group, founded by Trump. It was designed as an alternative to mainstream platforms, and it seeks to provide a space that advocates for free speech, particularly appealing to those who feel marginalized by established social media networks. Users can create profiles, post messages called "Truths," and engage in discussions, effectively mimicking the structure and function of platforms like X (previously Twitter).

The purpose of Truth Social is twofold: to offer a platform for conservatives and Trump supporters to share their views without fear of censorship and to provide Trump with a direct line of communication with his base. This strategy resonates with many who believe that mainstream networks disproportionately suppress right-leaning opinions. Despite its promotion as a "big tent" platform championing free expression, Truth Social has faced criticism for its content moderation practices, which tend to be stringent against dissenting views, particularly those critical of Trump or related political narratives (Associated Press, 2024).

The emergence of Truth Social has had a notable impact on the social media and political landscapes. By catering primarily to a conservative audience, it has contributed to the polarization of online discourse, creating echo chambers that reinforce partisan views. While mainstream platforms like Facebook and Twitter (now X) boast billions of users, Truth Social remains smaller (Associated Press, 2024).

As of January 2025, Truth Social has 6.3 million active users, a slight increase from the average of 5.9 million monthly users in 2024. The platform experienced notable fluctuations in user activity, peaking at 13.8 million users in March and dropping to a low of only 2.1 million in June. Since its launch, the Truth Social app has been downloaded approximately 8.5 million times across both the Google Play Store and the Apple App Store (Mortensen, 2025).

The most-followed accounts on Truth Social are (Mortensen, 2025)

- Donald J. Trump, who has 8.6 million followers. As the founder of the platform and a prominent political figure, Trump utilizes Truth Social to share his views and interact with his supporters.
- Donald Trump Jr., who shares insights on business and politics.
- Dan Bongino, a conservative commentator who ranks third and is known for his strong political opinions.
- Eric Trump, the Executive Vice President of the Trump Organization, whose posts are on the satirical site

- The Babylon Bee, which engages audiences with humorous takes on current events.
- Dinesh D'Souza, a conservative filmmaker and commentator.
- Sean Hannity, the Fox News host.

This diverse group of influential figures reflects Truth Social's appeal to a conservative audience and highlights its role in shaping political discourse.

An Analysis of Technology on Politics

Role of Social Media in Campaigns

Social media has fundamentally transformed political campaigning, shifting communication from one-way channels like TV and print to interactive, data-driven engagement. Candidates now speak directly to voters through platforms that allow real-time feedback, personalization, and amplification.

The 2024 presidential election highlighted just how central these platforms have become—especially among Trump voters and swing voters. A Navigator Research report revealed that:

- **52% of new Trump voters** and **45% of swing voters** relied primarily on social media for political news, compared to just 37% of the general electorate.
- Traditional broadcast news was less favored: only 31% of base Trump voters and 38% of swing voters used it as a primary source.

- Cable news, particularly **Fox News**, remained dominant among 52% of Trump's base.
- **Podcasts** also gained traction, with 21% of new Trump voters citing them as a key news source (Cousens, 2024).

These shifts signal a broader move toward alternative, digital-first media environments that shape voter perception and campaign strategy.

Campaigns increasingly employ data-driven micro-targeting, customizing messages based on user behavior, interests, and demographics. While this enhances efficiency and precision, it also raises ethical concerns, particularly around privacy, consent, and manipulation—as seen in the Cambridge Analytica scandal.

At the same time, social media has empowered grassroots movements and revolutionized fundraising. Movements like #MeToo demonstrate how digital platforms can mobilize support, drive public discourse, and exert political influence.

However, the digital shift also brings serious challenges:

- The spread of **misinformation** undermines informed decision-making and threatens **electoral integrity**, as seen in the 2016 U.S. election.
- **Algorithmic bias** fosters echo chambers, reinforcing polarization and limiting exposure to diverse perspectives.

Addressing these issues requires a multi-stakeholder approach. Tech companies, governments, and civil society must collaborate to improve transparency, enforce ethical data use, and boost digital literacy, helping voters navigate the complex, evolving landscape of political communication in the digital age.

Artificial Intelligence in Voter Analysis and Outreach

Technological advancements are changing politics and how we communicate. AI is now used to analyze voters' behavior and the effectiveness of campaigns. With AI, campaigns can create detailed profiles of voters and send messages that target specific groups. In the past, campaigns collected a lot of data by hand, but now politicians can quickly gather large amounts of information from social media, voting records, and online activities to understand voters better. This allows campaigns to focus on different issues for different groups, like urban professionals who care about climate change versus rural voters who are concerned about farming topics.

Artificial intelligence is rapidly reshaping political campaigns by optimizing advertising strategies through advanced analytics, enabling cost-effective targeting and real-time adjustments based on voter engagement. These tools allow campaigns to adapt messaging instantly, improving outreach precision and resource efficiency.

Accessible AI platforms also help level the playing field, empowering smaller or underfunded campaigns to reach wider audiences and compete with larger, better-funded opponents. AI-driven chatbots enhance direct voter interaction, providing instant responses and facilitating smoother communication between candidates and constituents.

However, the use of AI raises critical ethical concerns. Issues surrounding data privacy, algorithmic bias, and lack of transparency can erode public trust. To maintain the integrity of AI-driven campaigns, it is essential to implement clear policies that:

- Safeguard personal data,
- Ensure algorithmic fairness and inclusion,
- And promote transparency in how AI is used in voter outreach.

Maintaining these standards is vital to cultivating an informed, empowered electorate and protecting democratic values in the digital age.

Ethical Challenges of Artificial Intelligence Applications

Trump's Approach and Implications

President Trump's approach to AI emphasizes the importance of innovation free from ideological bias, as outlined in his recent Executive Order 14179. By revoking previous policies established during Biden's administration, Trump's order, with the purpose of eliminating barriers to AI development, asserts that any directive hindering progress will be suspended. Trump's executive order highlights the need for AI systems that promote "human flourishing, economic competitiveness, and national security" (The White House, 2025d, sect. 2), distancing his administration from what he perceives as politically motivated regulations.

Notably, Trump's order threatens to overturn safeguards designed to prevent AI discrimination based on race, gender, or disability, potentially undermining efforts to ensure fairness in technology. The executive order states (The White House, 2025d):

The United States has long been at the forefront of artificial intelligence (AI) innovation, driven by the strength of our free markets, world-class research institutions, and entrepreneurial spirit. To maintain this leadership, we must develop AI systems that are free from ideological bias or engineered social agendas. (sect. 1)

Executive Order 14179 includes a directive to create an AI action plan within 180 days (The White House, 2025d, sect. 4). Led by a team, including new advisor David Sacks, this action indicates an aggressive push toward deregulation in the AI sector. Critics worry that this change may prioritize the interests of tech billionaires over public safety and equity, raising concerns about the long-term impact on civil rights in the age of AI (Ochs et al., 2025).

The incorporation of AI in governance presents serious ethical dilemmas, particularly in relation to algorithmic bias. Such biases may produce unjust results, especially in critical decision-making scenarios such as recruitment and loan approvals, particularly when the historical data used for training reflects existing racial and gender biases. A significant additional issue is accountability; the opaque nature of AI makes it difficult to pinpoint responsibility in the event of mistakes. This situation calls for the development of legal frameworks that clarify liability and the responsibilities of developers, seeking to reduce the risks tied to AI implementation.

Privacy issues intensify due to AI's dependency on vast amounts of personal data, showing the need for strong data protection measures. Approaches like differential privacy can facilitate efficient AI functionality while protecting individuals' privacy rights. Moreover, the prospect of AI being used for surveillance incites concerns about government overreach and the potential infringement of civil liberties. Thoughtful policies are vital to finding a balance between ensuring security and preserving personal freedom.

AI Under Trump: Forward-Looking Observations

The following curated list of insights reflects common viewpoints from AI commentators, ethics researchers, and tech-forward thinkers—especially considering the current Trump administration's policy leanings and societal impact:

1. **Expect Deregulation to Accelerate Commercial AI Deployment**
   - Industry insiders anticipate looser federal constraints on AI applications in business, defense, and social platforms. Speed-to-market may outpace ethics reviews, which raises the need for independent watchdogs.
2. **Watch for National Security-Based Tech Bans**
   - The administration is likely to intensify scrutiny of Chinese companies (e.g. TikTok, Huawei, DJI), framing restrictions around data sovereignty and ideological influence. Expect more bans or forced divestitures.
3. **Federal Education May Prioritize "Skills-First" AI Literacy**
   - Policies might shift toward technical AI fluency in schools and vocational programs, possibly sidelining broader discussions on ethics and bias in favor of workforce preparation and competitiveness.

4. AI-Powered Surveillance Could Expand

- Civil liberty groups warn that AI facial recognition, predictive policing, and behavior tracking could be more widely embraced, especially under a law-and-order narrative.

5. Social Media Moderation May Loosen

- Commentators anticipate reduced pressure on platforms to moderate misinformation, which might open the door to AI-powered recommendation engines becoming more aggressive—and potentially more polarizing.

6. Public-Private AI Alliances Will Likely Intensify

- Expect more deals between federal agencies and companies like Palantir, OpenAI, or Amazon Web Services to deploy AI for national defense, immigration, and data analysis.

7. Decentralized AI Projects May Gain Popularity

- In response to centralized governance and potential ideological bias, developers may push harder toward open-source and decentralized machine learning architectures.

8. AI in Day-to-Day Living Will Become More Embedded

- From smart appliances to AI-driven scheduling and budgeting tools, expect further normalization of "quiet AI" that handles background tasks without fanfare—but with implications for data privacy.

9. Movement-Based AI Advocacy May Surge

- Similar to grassroots privacy and climate movements, AI watchdogs are predicted to organize more visibly around fairness, transparency, and algorithmic accountability under a populist administration.

10. Data Privacy Could Become a Personal Responsibility

- If federal protections weaken, consumers may need to rely more on encryption tools, browser extensions, and data literacy to safeguard their digital identities.

The Ethical Fault Lines in Trump's AI Agenda

In January 2025, President Trump issued the Executive Order on Removing Barriers to American Leadership in Artificial Intelligence, signaling a clear break from the ethics-centered framework of the previous administration. While the directive aimed to accelerate innovation and reduce regulatory friction, it also exposed significant ethical vulnerabilities in the nation's AI policy—raising concerns across civil rights, accountability, and international cooperation.

1. **Erosion of Oversight and Accountability:** By eliminating requirements for mandatory red-teaming of high-risk AI models and rolling back audit and incident-reporting directives, the administration removed key transparency mechanisms. Without enforced oversight, private developers now operate with minimal federal scrutiny—raising the risk of unintended harms, model misuse, and lack of public recourse.

2. **Neglect of Bias and Civil Rights Protections**: Unlike Biden's AI guidance, which included provisions to mitigate algorithmic discrimination, the new framework omits binding language on fairness, equity, and civil rights enforcement. While it vaguely affirms "trust and safety," it strips out references to protecting

marginalized communities—leaving critical gaps in protection where bias tends to hit hardest.

3. **Unilateralism and Global Ethical Divergence**: The executive order adopts an explicit "America First" posture, revoking collaborative efforts that previously aligned U.S. policy with global allies. This departure risks isolating the U.S. from international frameworks such as the EU's AI Act, weakening global cohesion, and eroding American moral leadership in responsible technology governance.

4. **Expansion of Surveillance and Predictive Policing**: Under the banner of national security and public safety, the order encourages expanded use of AI surveillance tools—from facial recognition to predictive policing—often without clear consent or community oversight. These tools have already raised concerns about privacy violations, profiling, and algorithmic injustice, particularly in under-resourced communities.

5. **Privatization of Ethical Responsibility**: With federal guardrails loosened, ethical accountability now falls to private companies. In a landscape driven by profit, this shift risks inconsistent standards and self-policing. Without uniform enforcement, the public may be left vulnerable to opaque decision-making and limited options for recourse when harms occur.

As the U.S. pivots toward deregulation in AI, the need for legislative checks, civil society engagement, and renewed public oversight becomes increasingly urgent. Innovation alone cannot justify the absence of ethical guardrails. Without firm commitments to transparency, fairness, and global cooperation, the pursuit of technological leadership may come at the cost of justice and public trust.

Bringing it All Together

- President Trump's executive orders have significantly reshaped the U.S. AI landscape, prioritizing deregulation and freedom of expression.
- Experts warn this approach may:
  - ▷ Normalize misinformation,
  - ▷ Complicate accountability,
  - ▷ And undermine public trust in democratic institutions.
- This chapter explored how AI-driven tools are transforming political campaigns and policies by enabling:
  - ▷ Microtargeting,
  - ▷ Optimized canvassing strategies,
  - ▷ Automated fundraising,
  - ▷ And highly personalized messaging at scale.
- While these innovations offer efficiency and precision, they raise serious concerns around:
  - ▷ Privacy,
  - ▷ Fairness,
  - ▷ And the power dynamics shaping voter engagement.
- As campaigns rely more heavily on AI and data analytics, democracy faces new questions:
  - ▷ How do we ensure transparency and informed consent?

- ▹ What rights should voters have over the use of their personal data?
- ▹ How can innovation be balanced with equity and accountability?

- The AI landscape is continually evolving on all fronts increasing efficiency through innovation and opportunity.
- The pace of change in the AI landscape often outpaces the associated Ethical considerations and responsibilities.
- Addressing these challenges is essential as digital strategy policies and actions and systems become increasingly connected.

Chapter 5

Regulatory and Financial Technologies—Regtech and Cryptocurrency on the Spot

Trump is significantly reshaping the financial market, particularly with his administration's pro-cryptocurrency policies that are revitalizing enthusiasm in the digital assets sector. Following his election for a second term, the price of Bitcoin surged to around $104,000, reflecting a 50% increase, as his administration swiftly enacted measures aligning with the desires of crypto advocates. Many in the industry had high hopes for Trump, anticipating a rollback of regulations perceived as barriers to crypto growth, particularly under former SEC Chair Gary Gensler. The recent appointment of former SEC Commissioner Paul Atkins and the installation of venture capitalist David Sacks as the White House AI and crypto czar signal a broader commitment to developing a favorable crypto environment (Ermey, 2025).

One striking development is the launch of Trump's own meme coin shortly before his inauguration which underscored a personal investment in the cryptocurrency trend. This initiative, alongside First Lady Melania Trump's own coin, signifies a bold endorsement of digital assets. While the volatility of meme coins raises skepticism among retail traders, experts interpret this as a vote of confidence in the crypto ecosystem. Overall, Trump's administration is poised to position the US as a "Bitcoin superpower." The government has emphasized an approach toward cryptocurrency as a groundbreaking strategy to ensure economic progress.

While cryptocurrency revolutionizes the financial market, regulatory and financial technologies are redefining compliance and innovation regarding all economic activities related to the technology industry. Regtech, a subset of financial technology regulation, is at the forefront of these transformations at the management level. In our current economic landscape, traditional compliance measures are outpaced by rapid developments, and the different stakeholders need to adapt and adopt tech-infused approaches to maintain order and efficiency.

As companies become increasingly reliant on AI and automation tools, regtech emerges as a game-changer, streamlining regulatory processes and vastly reducing the associated costs. Let's explore how these technological advancements are reshaping the interaction between regulation and finance while we examine the Trump administration's AI action plan and recent policy shifts.

Cryptocurrency and New Policy

Cryptocurrency has emerged as a transformative force in the global financial landscape, reshaping how individuals and institutions interact with money and investments. As digital assets gain mainstream acceptance, their importance is increasingly highlighted not only in economic terms but also in political and governance contexts. The rise of cryptocurrencies, such as Bitcoin and Ethereum, has prompted discussions about their potential to democratize finance, enhance transaction efficiency, and challenge traditional banking systems.

Countries around the world have taken measures to accompany this evolution in the past few years. However, policymakers still grapple with how to integrate these emerging technologies into existing legal frameworks while addressing concerns related to security, taxation, and consumer protection.

Moreover, the intersection of cryptocurrency with politics has become increasingly evident as political figures and movements leverage the technology to mobilize support and engage with constituents. In this environment, cryptocurrencies have sparked debates around issues like financial sovereignty, privacy, and the implications for state control over monetary policy. As such, understanding the multifaceted dynamics of cryptocurrency—its economic significance, regulatory challenges, and role in political discourse—is essential for navigating the future of governance in an increasingly digital economy. For President Trump, cryptocurrency stands in the center of the scene.

Trump's Cryptocurrency: A Family Memecoin

President Trump has notably embraced cryptocurrency, signaling a significant shift in his approach to digital assets. On January 17, 2025, he announced the launch of the $TRUMP meme coin, which rapidly surged in popularity—reaching a peak market value of over $14.5 billion before experiencing a sharp decline. Despite the volatility, the coin reportedly generated between $86 million and $100 million in trading fees, highlighting the lucrative potential of meme-based crypto ventures—even as many small traders suffered financial losses (Wilson & Conlin, 2025).

What Are Meme Coins?

A meme coin is a type of cryptocurrency inspired by internet memes or pop culture, created largely for speculation rather than functionality. Unlike traditional digital assets like Bitcoin or Ethereum, which are designed around advanced blockchain utilities or security features, meme coins tend to lack intrinsic value. Their appeal is largely fueled by community engagement, online buzz, and celebrity endorsement—making them highly volatile and often associated with pump-and-dump schemes.

A Family Affair in Crypto

The $TRUMP token features an image of Donald Trump and was soon followed by a parallel release: First Lady Melania Trump's own meme coin, which briefly peaked at over $13 before stabilizing near $2.30 (Ermey, 2025). These ventures sparked widespread discussion about the Trump family's growing footprint in the meme coin market. Some interpreted these actions as a sign of the administration's support for crypto-friendly policy, while others voiced concern over the speculative and potentially manipulative nature of such coins (Wilson & Conlin, 2025).

Meme Coins as Cultural Currency

These tokens quickly gained traction across social media platforms and online communities, generating a buzz that extended well beyond conventional fundraising methods. Their success reveals cryptocurrency's power to function as more than just a financial asset—it has become a symbolic tool for mobilization, signaling loyalty, and expressing identity among political supporters.

This trend reflects a broader shift in how technology, branding, and politics intersect. Crypto assets like meme coins are no longer separate from campaign culture—they are now a part of it, blending finance with community-building, identity signaling, and digital activism.

Unlike Bitcoin, Ethereum and others which derive value from technical innovation or secure infrastructure, meme coins are driven almost entirely by cultural momentum. Their prices are sensitive to viral trends, celebrity involvement, and social media buzz. As such, they can deliver dramatic returns—but just as easily, steep losses. This kind of volatility makes them both enticing and dangerous, especially for inexperienced investors (Krause, 2025).

Politically themed meme coins—like $TRUMP and $MELANIA—illustrate the evolving role of cryptocurrency in campaign branding, speculation, and potentially illicit fundraising. Some governance tokens, such as WLFI, introduce additional complexity, especially when their financial structures intersect with those of politically exposed persons (PEPs). These examples underscore why regulatory agencies are ramping up scrutiny around transparency, conflict of interest, and the ethical use of digital assets in public life.

Rising Red Flags and Regulatory Gaps

Recent research highlights a growing list of concerns tied to politically affiliated meme coins, including:

- **Market manipulation**,
- **Insider trading**,
- **Concentrated ownership**,
- And **regulatory loopholes** that allow public figures to benefit from crypto schemes without proper disclosure.

Scenarios involving high-profile individuals funneling funds through affiliated tokens present clear conflicts of interest, eroding investor trust and raising questions about political accountability (Krause, 2025).

As a result, calls are intensifying for:

- **Stronger oversight**,
- **Integration of securities laws**,
- And the creation of **ethical standards** tailored for digital political finance.

Regulators are increasingly advocating for a harmonized approach, combining traditional financial safeguards with ethical frameworks, to close gaps in market integrity and prevent the misuse of cryptocurrency for personal or political gain.

Other Policies

Trump's administration has passed other regulatory reforms that embrace a crypto-friendly stance. One of the most notable outcomes of this approach has been the dismissal of federal investigations and lawsuits against major cryptocurrency platforms. In February 2025, a federal judge approved a request from the SEC and Binance to pause their litigation while awaiting new regulations from the Crypto Task Force. The SEC has also sought similar pauses in other cases, including one against Justin Sun, a Chinese crypto entrepreneur. Notably, the SEC has dropped its lawsuit against Coinbase and has cleared other platforms, such as Robinhood, Uniswap, OpenSea, and Consensys, from investigations related to their crypto activities (Khalili, 2025).

This signals a more lenient regulatory environment that allows these firms to operate with greater freedom, raising a sense of security among investors and industry stakeholders. By reducing the threat of legal challenges, the administration effectively encourages innovation and participation in the cryptocurrency market, accelerating the sector's growth and legitimacy.

Additionally, recent revisions by the Federal Deposit Insurance Corporation (FDIC) have empowered banks to engage in certain cryptocurrency activities without needing prior approval. This policy update marks the adherence to a different paradigm in the relationship between traditional financial institutions and the evolving crypto space. Enabling banks to compete with established cryptocurrency firms opens avenues for collaboration and innovation, facilitating the development of new financial products and services that incorporate digital assets. This increased participation from banks not only legitimizes cryptocurrencies but also enhances overall market stability, as trusted financial institutions provide a layer of oversight and security.

Moreover, the ability of banks to engage in crypto activities positions them to leverage their existing infrastructure and customer base to introduce cryptocurrencies to mainstream users. This can potentially lead to wider adoption, as consumers may feel more comfortable investing in cryptocurrencies through established banking channels. The combined effect of dismissing federal cases and allowing bank participation in crypto activities has created a more favorable regulatory environment, encouraging innovation while simplifying compliance for businesses operating in the space.

Regulation and Challenges

The decentralized nature of digital currencies undermines centralized control over financial transactions, compelling governments to reconsider regulatory frameworks for maintaining economic stability. Cryptocurrencies gain traction and test the resilience of traditional banks as blockchain technology introduces new paradigms for financial inclusion (Safarli & Safarli, 2024).

However, the volatile nature of cryptocurrencies poses unknown risks. Price fluctuations can create financial instability, impacting both investors and broader economies. Thus, a reevaluation of the regulatory frameworks that might be needed in upcoming environments is required (Scicchitano, 2020).

Moreover, cryptocurrencies are often linked to illicit activities due to their pseudonymous nature. To limit the space for illegal activities, law enforcement, financial institutions, and tech collaborate to provide essential enhanced security and AML strategies. Simultaneously, it is equally important to raise public awareness and inform users about potential risks associated with digital currencies (Safarli & Safarli, 2024).

The $LIBRA cryptocurrency scandal, for instance, is evidence of the precarious intersection of politics and cryptocurrency. Argentine President Javier Milei's involvement introduces questions about the roles political leaders should or should not assume. On February 14, 2024, Milei promoted $LIBRA on social media, which resulted in a rapid price surge from $0.000001 to $5.20, but following his withdrawal of support, the currency plummeted to $0.99 within hours. While $LIBRA's market capitalization peaked at $4.6 billion, it ultimately fell to just $162 million, leading to significant financial losses for hundreds of investors, many of whom were reportedly from Argentina and the US (Nicas & Yaffe-Bellany, 2025).

The situation escalated as accusations of insider trading and "pump and dump" schemes emerged. These accusations led to a lawsuit against Milei for various allegations, including fraud. The scandal has profoundly shaken public trust in Milei's government, jeopardizing Argentina's credibility in the international arena and complicating relationships with financial institutions like the International Monetary Fund (IMF) (Nicas & Yaffe-Bellany, 2025).

Interestingly, figures like Donald Trump and Elon Musk have expressed support for Milei, reflecting a broader trend where political endorsements of cryptocurrencies can precipitate both rapid financial gains and acute regulatory scrutiny. The $LIBRA case exemplifies the risks associated with meme coins, often prone to volatility and speculation, severely impacting investor confidence and governance structures in emerging digital economies.

The cryptocurrency market is heavily influenced by prominent figures in the tech industry and politics. Underlining their power in shaping public perception and acceptance of digital assets is crucial. Their endorsements can drive massive market fluctuations, often leading to both excitement and confusion among investors. Moreover, the intertwining of influential political figures and tech entrepreneurs complicates the regulatory landscape.

Trump's Policies Toward Cryptocurrency Markets

With a renewed commitment to establishing the US as the "crypto capital of the planet," Trump launched Executive Order 14178, which we have explored in previous chapters (Shivaram, 2025). With the objective of fostering the development and responsible management of digital assets., a key element of this initiative is the formation of the President's Working Group on Digital Asset Markets. This group, tasked with proposing new regulations for cryptocurrencies and exploring the creation of a national cryptocurrency stockpile, highlights a shift away from previous restrictive regulatory frameworks.

Overall, Trump's policies represent a concerted effort to bolster the cryptocurrency market, positioning it at the intersection of technology, finance, and governance amid an increasingly competitive global landscape. The administration's favorable stance on cryptocurrencies aims to stimulate investment and engagement in the digital asset space while simultaneously navigating the complex regulatory environment that accompanies this transformative technology.

The Market's Response

Trump has positioned himself as the "crypto president," expressing a commitment to making the US a leader in cryptocurrency adoption. This has been evident in his appointments of crypto-friendly figures, such as Paul Atkins as chairman of the SEC, setting a more lenient regulatory environment for the cryptocurrency industry.

Amid these changes, figures like Tim Cook from Apple, Jeff Bezos from Amazon, Sundar Pichai from Google, Mark Zuckerberg from Meta, Sam Altman from OpenAI, and even TikTok's Shou Zi Chew have also bolstered the administration's crypto-friendly image, aligning themselves with Trump's agenda. Moreover, all of them scheduled meetings with the president to engage in the new policies (Kerr & Bhuiyan, 2025). The involvement of individuals from the tech industry, which has historically faced scrutiny under stringent regulations, highlights the intertwining of cryptocurrency and political interests.

As a result of this new approach in policy, the cryptocurrency market witnessed a notable rally, with prices of major assets, including Bitcoin, reaching new highs amid increased institutional interest and important regulatory updates that promised to ease compliance burdens. This surge was characterized by a heightened amount of trading activity, as confidence in the market was restored with the prospect of a more lenient regulatory environment.

However, the market also showed signs of volatility typical of cryptocurrencies, with some assets, such as meme coins linked to Trump's presidency, experiencing wild price swings. The speculative nature of these coins attracted new investors eager to capitalize on potential gains while simultaneously raising concerns about the sustainability of such trends.

Overall, Trump's regulatory changes and the resultant adaptive response in market sentiment created an environment ripe for both innovation and speculation. Investors responded swiftly to the perceived opportunities, illustrating the intricate relationship between political actions and cryptocurrency market dynamics. While optimism prevailed, the unpredictable nature of cryptocurrencies continued to remind participants of the inherent risks involved in this evolving financial landscape (Kerr & Bhuiyan, 2025).

Regulatory Uncertainty and Compliance Challenges

Regulatory uncertainty is a major challenge facing the rapidly growing field of cryptocurrency. The decentralized and global nature of cryptocurrencies complicates efforts to establish a consistent regulatory framework, resulting in confusion among market participants. Each jurisdiction has its interpretation of laws, leading to vastly different legal environments that make compliance difficult for crypto businesses. For example, investment advisers navigating the "Custody Rule" under the Investment Advisers Act of 1940 face ambiguity regarding whether crypto custodians meet required qualifications, which can hinder institutional participation in crypto markets (Adams et al., 2025).

Disparities in regulations concerning the classification of crypto assets as securities or commodities further complicate compliance for international businesses. Some countries impose stringent regulations on cryptocurrencies as securities, while others treat them as commodities, leading to lighter oversight. This inconsistency breeds operational inefficiencies and stifles innovation, as entrepreneurs may hesitate to undertake new projects without clear legal guidelines. Moreover, varying regulatory environments create barriers to seamless global operations, complicating cross-border transactions essential for the crypto industry's success.

Despite these obstacles, initiatives like the SEC's Crypto Task Force aim to clarify regulations by establishing clearer registration pathways and enhancing disclosure frameworks. By supporting innovation while ensuring investor protection, such efforts could provide the much-needed clarity to bolster confidence among market participants and bridge the gap between regulation and innovation.

Balancing Innovation With Effective Oversight

Balancing innovation with effective oversight in the rapidly evolving cryptocurrency landscape is essential under President Trump's administration. As cryptocurrencies increasingly shape financial systems, their decentralized nature creates a pressing need for structured regulatory frameworks that safeguard consumers without stifling technological progress (Alto, 2024). U.S. regulations must strike a balance—creating environments conducive to innovation while implementing necessary protections against risks like fraud and market manipulation.

Excessive regulations can deter entrepreneurs and inhibit progress, as evidenced by countries like China, which has imposed strict regulations, effectively banning cryptocurrency trading and ICOs (Alto, 2024). Conversely, insufficient regulation can expose the market to vulnerabilities, opening doors for scams and fraudulent schemes that target uninformed investors. Hence, the US aims to cultivate a balanced regulatory approach that protects investors without hampering innovation (Adams et al., 2025).

The recent appointments in the SEC, such as the nomination of Paul S. Atkins as chairman and the establishment of the Crypto Task Force led by Hester Peirce, adopt a scope for a more innovation-friendly regulatory environment. This task force's objective is to establish clear regulatory frameworks and practical registration pathways that will promote responsible innovation.

International cooperation will also play a critical role in addressing the complexities of cryptocurrency regulations. Given the global nature of digital assets, consistent regulations across jurisdictions can enhance market stability and boost investor confidence (Alto, 2024). As regulatory authorities refine their approaches, ongoing public education about the risks and benefits of cryptocurrencies can empower investors, contributing to a safer, more innovative crypto landscape where technological advancement and consumer protection can coexist harmoniously.

Global Coordination in Cryptocurrency Regulation

The borderless characteristic of cryptocurrencies means that regulatory bodies across countries must work together to establish coherent laws that address the complexities of decentralized finance. Without this cooperation, the risk of a fragmented regulatory landscape arises, where varying rules create loopholes that bad actors can exploit (Adrian et al., 2023). Consistent standards and guidelines mitigate risks related to regulatory arbitrage and promote a safer investment environment, with the principle of "same activity, same risk, same regulations" serving as a foundational strategy (Kortam, 2022).

Furthermore, harmonized regulations enhance market stability and foster investor confidence. Clear and predictable rules allow investors to make informed decisions, which is crucial for participation in the crypto market. This predictability can drive consumer trust and market order (Kortam, 2022).

Despite these advantages, achieving an effective global regulatory framework is challenging due to differing national priorities and economic considerations. Balancing the interests of consumer protection and innovation requires diplomacy and negotiation among international stakeholders. Institutions like the FSB and the International Organization of Securities Commissions (IOSCO) play vital roles in guiding these efforts. To overcome existing obstacles, strong coordination at domestic and international levels, along with public-private partnerships, is critical for shaping practical regulations that effectively address the evolving crypto landscape.

Trump in the Global Landscape

The U.S. policy shift under Trump's administration regarding cryptocurrencies has far-reaching global implications, altering the dynamics of international regulatory frameworks and influencing market behavior worldwide. As the US adopts a more supportive stance on digital assets, countries may feel compelled to reevaluate their regulatory approaches in order to remain competitive in the rapidly evolving cryptocurrency space. This potential reevaluation could lead to a wave of countries adopting more favorable regulations that encourage innovation and attract investment, thereby enhancing their positions in the global market.

However, the cryptocurrency environment remains highly volatile, significantly impacted by external factors such as trade policies. For instance, recent tariff threats had a pronounced effect on market sentiment, resulting in a sharp downturn where Bitcoin's value sank below $80,000, and the total market capitalization of cryptocurrencies experienced a loss exceeding $1 trillion since December (Kerr & Bhuiyan, 2025).

Such fluctuations underscore the intricate interplay between regulatory support and macroeconomic factors, illustrating how trade tensions and geopolitical developments can deeply affect investor confidence and market stability.

As the market continues to wrestle with these challenges, developments such as the White House Crypto Summit are pivotal in shaping the direction of U.S. cryptocurrency policy.

The White House Crypto Summit

On March 7, 2025, U.S. President Donald Trump hosted the first-ever White House Crypto Summit as an attempt to establish the US as a global leader in digital financial technology and address several key topics critical to the future of the cryptocurrency landscape. Prominent figures from the cryptocurrency industry, such as Zach Witkoff, cofounder of World Liberty Financial, Michael Saylor, CEO of MicroStrategy, Brian Armstrong, cofounder and CEO of Coinbase, investors Cameron and Tyler Winklevoss, entrepreneur David Bailey, and David Sacks, the White House crypto czar attended to discuss Trump's plans for a government-owned stockpile of digital assets.

Central to the summit's objectives was the discussion surrounding regulatory clarity. Participants explored the establishment of clear and supportive regulatory frameworks designed to encourage innovation within the crypto industry. This included examining how digital assets can be more effectively integrated into the broader financial system to enhance economic opportunities.

Expected outcomes from the summit included shaping future regulatory approaches for digital assets, including discussions on stablecoins and the potential establishment of Bitcoin reserves. Additionally, the summit was geared toward promoting innovation in the cryptocurrency space while ensuring consumer protection and financial stability. However, some of the summit results deferred from expectations.

A key outcome of the summit was the announcement of an executive order to establish a strategic reserve of Bitcoin, alongside plans for a digital asset stockpile that would include other cryptocurrencies like Ethereum and Ripple, all while ensuring no additional costs to taxpayers.

Despite the significant gathering and Trump's ambitions to position the US as a leader in cryptocurrency, many attendees left disappointed due to the lack of concrete plans for acquiring new digital assets, leading to a decline in Bitcoin prices following the event.

A Bitcoin Reserve

Aligned with the topic and outcomes of the summit, President Donald Trump launched Executive Order 14233, *Strategic Bitcoin Reserve and United States Digital Asset Stockpile.* The purpose of this order is to establish the "Strategic Bitcoin Reserve" using cryptocurrencies already owned by the federal government, primarily those seized through criminal or civil asset forfeiture. This decision disappointed some market participants who had anticipated a structured plan for purchasing new Bitcoin tokens. Following this announcement, the price of bitcoin dropped approximately 5% to $85,000, although it later rebounded to around $89,200 during early European trading (Bose & Ward, 2025).

The initiative, led by the White House crypto czar David Sacks, opens the door for potential future acquisitions of Bitcoin. It allows the U.S. Commerce and Treasury Secretaries to develop budget-neutral strategies for acquiring additional Bitcoin, which ensures no extra costs to American taxpayers.

Experts noted that the significance of this executive order lies mainly in its symbolic nature, marking the first formal recognition of Bitcoin as a reserve asset by the U.S. government. However, it explicitly states that the reserve will only consist of bitcoin already held by the government, with no clear indication of future acquisitions or timelines. The release of this order seemed to fall short of expectations, particularly after a previous surge in bitcoin prices to an all-time high earlier this year, leading some to express disappointment over what they perceived as a lack of active buying, equating the reserve to "a pig in lipstick" (Bose & Ward, 2025).

Understanding Regtech and Its Applications

In the fast-evolving financial landscape, the integration of regulatory technologies (regtech) is becoming a critical necessity. As companies and institutions face increasing regulatory demands, streamlining compliance processes through regtech offers notable advantages in efficiency and cost-effectiveness. Traditional methods of compliance can be cumbersome and resource-intensive, leading to significant operational costs. Regtech solutions leverage technological advancements to automate these processes, allowing firms to reduce manual labor and focus their efforts on core activities. This not only lowers costs but also enhances the precision and speed with which firms can respond to regulatory requirements.

Emerging technologies, particularly AI, play a pivotal role in the development of regtech applications. For instance, by automating transaction monitoring and reporting tasks, AI-driven regtech tools can efficiently handle large volumes of regulatory information, ensuring firms stay aligned with the latest standards without delay. The agility of these technologies allows financial institutions to remain compliant while also embracing innovation, helping them maintain a competitive edge.

An example of a specific regtech solution is "ComplyAdvantage," which offers AI-driven insights to help businesses manage their compliance risk related to AML regulations. It aggregates data from various sources to provide real-time risk assessments, enabling organizations to stay ahead of regulatory changes and ensure they are meeting their legal obligations. This not only enhances efficiency but also reduces the likelihood of costly compliance failures or penalties.

Looking toward the future, several trends are poised to enhance regtech's impact on the financial sector further. Advancements in machine learning are set to refine the predictive analytics abilities of these technologies, offering even greater foresight into potential compliance issues before they arise. Moreover, as regulatory bodies become more familiar and comfortable with regtech applications, there's a growing likelihood that these technologies will be officially embraced, transforming them into standard components of regulatory practices globally. This endorsement could lead to more coherent and streamlined regulatory frameworks, reducing the complexity currently faced by financial entities navigating diverse international standards.

Furthermore, as the global financial environment becomes increasingly interconnected, the harmonization of international regulations through technology-driven solutions will likely gain momentum. This may lead to joint efforts supported by interconnected technological systems to establish uniform compliance standards internationally. In this context, regtech could serve as a unifying force, enabling different jurisdictions to coordinate effectively, thus enhancing overall market stability.

Impact on Financial Regulations

The swift pace of technological advancement demands a reevaluation of traditional oversight methods. Regulators must adapt to new realities where financial transactions occur in mere nanoseconds and large volumes of data flow through complex digital channels. This transformation necessitates reconsidering existing frameworks and exploring innovative strategies that can effectively oversee modern fintech operations. Regtech has emerged as a pivotal tool in this space, assisting companies in automating compliance tasks and ensuring timely adherence to regulatory changes (Haddad, 2025).

However, the integration of technology into compliance frameworks presents its own set of challenges. Chief among them is the ongoing effort to maintain consumer protection amid the complexities introduced by fintech advancements and global regulatory variations. As fintech companies expand their reach across borders, they encounter heterogeneous regulatory environments, leading to challenges in uniform implementation (Haddad, 2025).

The absence of standardized global regulations causes confusion and increases operational costs for firms attempting to comply with diverse standards. Fintech companies must, therefore, be diligent in understanding and adhering to the specific regulations of each jurisdiction in which they operate. This multijurisdictional requirement often means adapting compliance programs to manage cross-border transactions effectively.

A notable case of a U.S. fintech company facing challenges with regtech implementation is ZoomInfo Technologies. In 2024, as the company aimed to enhance automation within its data privacy processes to comply with regulations such as the California Privacy Rights Act (CPRA) and Europe's GDPR, it encountered significant internal resistance. Specifically, ZoomInfo's security and finance departments exhibited hesitancy toward adopting solutions from emerging start-ups, raising concerns about the reliability and security of these relatively unproven vendors (Vartebedian, 2024).

Despite these obstacles, ZoomInfo moved forward with implementing the regtech solutions, instituting additional safeguards, such as restricting start-up vendors' access to their data. This strategic approach enabled the company to automate certain compliance tasks, thus enhancing operational efficiency without the need for hiring additional full-time employees.

Moreover, fintech companies and regulators are compelled to collaborate for informed policymaking. In an industry characterized by rapid innovation, having open communication ensures that regulations keep pace with technological developments while preserving market integrity. Shared efforts can bring an understanding of technology's potential and risks and provide the guidelines for policies that are both effective and flexible enough to accommodate future innovations. For instance, fintech firms can provide valuable insights into emerging trends and technologies so regulators can craft relevant and forward-looking policies. Similarly, regulators can assist fintech companies' performance in legal scenarios to ensure that new products and services meet necessary standards from inception.

Trump's Administration and Regtech

President Trump's policies regarding regtech have initiated significant changes in the financial sector, reflecting his broader deregulatory agenda. His administration's recent discussions about restructuring major regulatory institutions, such as the FDIC and OCC, aim to streamline regulatory frameworks. While some in the banking community view this as an opportunity for efficiency, there are concerns about the effect on financial oversight. The potential consolidation of agencies could create uncertainty in regulations and reporting requirements, which regtech firms need to navigate.

Moreover, Trump's administration emphasizes a pro-crypto stance alongside a commitment to reducing regulatory burdens. This includes revoking previous protections intended to curb misinformation in the marketplace, such as eliminating barriers pointed out as biased restrictions to free expression. The anticipated increase in demand for automated compliance solutions reflects the move towards a more streamlined regulatory environment where standardization might become necessary. However, with reduced federal oversight, state regulators may rise in prominence, necessitating further adaptation from regtech firms.

Looking Ahead: 10 Expert Insights on Trump-Era Crypto Policy

With the Trump administration's assertive embrace of cryptocurrency—through deregulation, the signing of the GENIUS Act, and the creation of a Strategic Bitcoin Reserve—the tone among crypto analysts has shifted. While enthusiasm is rising in some circles, it's tempered by questions about ethics, market stability, and long-term implications.

The following ten insights reflect the prevailing sentiment among experts and commentators navigating this new policy environment:

1. **Regulatory Clarity May Accelerate Innovation**: The GENIUS Act, now enacted, introduces clearer federal guidelines for stablecoin issuers. Many believe this will unlock broader mainstream adoption and pave the way for institutional capital to enter the market.

2. **Institutional Capital Could Flow In**: Policy support for crypto-friendly retirement portfolios, including 401(k) access to digital assets, could direct billions into the sector, fundamentally altering its scale and investor profile.

3. **Stablecoins Positioned to Reinforce Dollar Dominance**: New rules requiring full reserve backing (in dollars or Treasuries) position stablecoins not just as tools of innovation, but as instruments to strengthen the U.S. dollar's global role.

4. **Bitcoin Reserve May Tighten Supply**: By retaining seized Bitcoin in a strategic reserve rather than auctioning it, the administration may reduce available supply—potentially exerting upward pressure on market prices.

5. **CBDCs Are Off the Table—for Now**: The administration has banned federal development of central bank digital currencies (CBDCs), citing concerns over privacy, sovereignty, and potential misuse of centralized financial control.

6. **U.S.-Based Crypto Firms May Gain a Competitive Edge**: With regulatory backing and political alignment, domestic crypto ventures are likely to gain momentum and outpace foreign competitors—at least in the short term.

7. **Transparency Concerns Linger Around Political Ties**: Ventures linked to Trump-aligned figures, such as the USD1 stablecoin, have drawn scrutiny. Experts warn that favorable policies could lead to conflicts of interest and diminished market trust.

8. **DeFi Platforms May Face Looser Oversight**: Under the leadership of figures like SEC Chair Paul Atkins, enforcement actions are reportedly being scaled back—signaling a more hands-off stance toward decentralized finance (DeFi) ecosystems.

9. **Volatility Will Persist Around Politicized Tokens**: Politically branded tokens like $TRUMP continue to demonstrate boom-and-bust cycles. Analysts caution against conflating political loyalty with sound financial fundamentals.

10. **Legislative Momentum Will Likely Continue**: With multiple bills passed during the administration's self-declared "crypto week," further regulatory reform appears imminent—especially around market oversight and consumer protection.

Together, these developments suggest that while the Trump administration is steering crypto policy toward rapid growth and deregulation, the path forward will demand careful scrutiny to ensure innovation doesn't outpace accountability.

Future Implications of These Crypto Insights

While risks remain, many experts view this moment as an inflection point for the maturing digital asset ecosystem.

Market Growth and Structural Maturation

As stablecoin standards become codified, barriers to entry will drop, enabling a surge of new issuers and liquidity networks. Institutional-grade infrastructure—featuring custodial safeguards, audited reserve reports, and insured wallets—is expected to become the baseline. In this environment, compliant, dollar-backed tokens may gradually displace loosely regulated alternatives, nudging decentralized finance (DeFi) platforms toward models centered on transparency and regulatory alignment.

Institutional and Retail Capital Inflows

With retirement portfolios like 401(k)s gaining crypto access, the stage is set for unprecedented capital inflows from both institutional and retail investors. Insurance providers and asset managers are preparing specialized crypto offerings—ETFs, tokenized debt products, and yield-generating vaults—designed to attract long-term capital. On the retail side, integration into mainstream brokerage platforms will normalize digital assets as part of everyday financial portfolios, accelerating public adoption.

Geopolitics and Dollar-Dominance Dynamics

Stablecoins are poised to reinforce U.S. dollar dominance in global trade. As multinational corporations and sovereigns increasingly adopt dollar-pegged tokens for cross-border transactions, alternative regimes like the digital euro and China's e-CNY will face growing pressure to adapt or lose relevance. In emerging markets, demand for onshore dollar liquidity will rise as local players turn to U.S.-backed stablecoins to hedge against currency instability.

Supply Dynamics and Price Structures

The administration's decision to retain government-seized Bitcoin—rather than auction it—introduces a new dimension to market scarcity. By limiting circulating supply, this policy may contribute to sustained bullish sentiment and drive a broader divergence between spot and futures pricing. Derivatives markets are expected to respond with innovation, offering reserve-backed instruments, lending products, and structured yield strategies. For retail investors, higher Bitcoin prices could redirect capital toward altcoins and DeFi protocols offering competitive returns.

Innovation in Decentralized Finance

Reduced enforcement from agencies like the SEC is likely to trigger a wave of experimentation across DeFi. New lending models, tokenized real-world assets, and self-custodial insurance products will emerge. Developers are expected to prioritize compliant cross-chain bridges and Layer-2 solutions, making institutional onboarding more seamless. This convergence of open-source development and venture capital interest will fuel the next generation of decentralized financial infrastructure.

Transparency, Ethics, and Public Trust

As political figures become more entwined with crypto ventures, calls for ethical oversight will intensify. Projects with public affiliations may face new disclosure mandates, while third-party auditors and emerging self-regulatory bodies could play a larger role in certifying ethical standards. Trust in crypto will increasingly rely on verifiable on-chain data, clear governance structures, and a demonstrable commitment to transparency.

Legislative and Policy Evolution

Building on the GENIUS Act, lawmakers are expected to push forward legislation addressing crypto taxation, consumer protections, and interoperability standards for digital assets. States will compete to position themselves as crypto innovation hubs, offering favorable licensing regimes and tax incentives. Internationally, the U.S. framework may serve as a template for convergence, with global bodies like the G20 and BIS working toward harmonized regulatory approaches.

Bringing it All Together

- As the cryptocurrency sector evolves, the integration of regulatory technology (regtech) is becoming crucial for companies navigating heightened compliance expectations.
- The Trump administration's lenient regulatory stance may prompt global reconsideration of crypto policy, influencing international norms and potentially shifting the balance of market power.
- This deregulatory approach introduces both opportunities and risks—stimulating growth, but also raising concerns about market integrity, transparency, and power concentration in private hands.

- The key challenge ahead lies in crafting frameworks that support innovation while upholding ethical standards and protecting public interest.
- Policy, power, and decentralization intersect, it is important to consider what's at stake if regulation falls too far behind the pace of change.

Chapter 6

Transparency and Accountability in Governance

Transparency and accountability are pivotal components of effective governance, shaping the trust between government institutions and the public. With the rise of digital tools and platforms, the potential for open access to information and citizen engagement has expanded exponentially. This shift offers both opportunities and challenges as governments strive to balance transparency with security and ethical concerns. The intersection of technology and governance invites a deeper exploration into how these elements collectively influence democratic processes and public participation.

On January 20, 2025, Trump passed Executive Order 14147, *Ending the Weaponization of the Federal Government*, which addresses the perceived misuse of federal law enforcement and intelligence agencies against political opponents during the previous administration. The objectives of this order include ensuring accountability for actions described as targeting individuals who opposed the prior administration, including investigations and prosecutions that seemed politically motivated rather than based on legitimate legal grounds (The White House, 2025a, sect. 1).

However, this order raises concerns about transparency and accountability in government actions. By revoking security clearances for approximately four dozen former national security officials who questioned the integrity of his prior administration's actions, Trump creates a climate where dissent is stifled. This environment heightens fears about the potential misuse of AI and other technologies in government, as the lack of accountability may enable biased or politically motivated enforcement actions. The absence of checks and balances can lead to questionable applications of AI, compromising citizens' rights and undermining public trust in governmental processes.

Technological Tools for Transparency

Technology has emerged as a powerful ally that can be used to enhance transparency and accountability or to obscure processes. However, when we discuss technology in relation to governance, we must address both opportunities and concerns.

Open Data Initiatives

The advent of big data analytics has revolutionized governance by enabling data-driven decision-making that aligns policies with public needs and preferences. Governments can now analyze vast datasets to predict trends and tailor initiatives, allowing for proactive rather than reactive strategies. For example, demographic and economic data can help policymakers anticipate changes in employment or education needs, ensuring the effective allocation of public resources.

Also, by utilizing big data, governments enhance efficiency in service delivery, as seen in transportation planning or healthcare resource allocation, ensuring that citizens receive timely services. Governments can leverage AI in various ways to enhance their services. For instance, the Queensland Government in Australia employs machine learning and computer vision to automatically map and classify land use features from satellite imagery. This reduces the costs associated with land use mapping and enables more effective responses to biosecurity threats and natural disaster events. Regarding healthcare services, during the COVID-19 pandemic, Norway's Labor and Welfare Administration utilized a conversational AI named Frida to assist citizens in accessing social benefits around the clock. Frida successfully resolved 80% of inquiries without the need for civil servant intervention (*Governing With Artificial Intelligence*, 2024).

However, the integration of big data into governance raises significant concerns about privacy and ethics. The extensive volume of data collected risks infringing on individual privacy rights if not managed properly, underscoring the need for robust data protection laws. Existing regulations, like the Revised Common Rule, often inadequately shield publicly available data from misuse. Implementing strong privacy protections is crucial to building public trust, as citizens are more likely to support data initiatives when assured of their information's safety.

Additionally, there are ethical considerations regarding biases in algorithms that can lead to unfair outcomes. Governments must embed safeguards to identify and correct these biases and conduct regular audits of algorithmic performance. Real-time monitoring capabilities provided by big data enhance governmental transparency and accountability, enabling swift responses to emerging issues. By establishing clear guidelines for data sharing and fostering collaborations with technology and ethics experts, governments can maximize the benefits of big data while serving the public interest (Schubert & Barrett, 2024).

Blockchain Technology

In the financial realm, blockchain technology is known for its secure and immutable nature. Blockchain provides a robust framework for recording government transactions. Once a transaction is entered into the blockchain, it cannot be altered, ensuring traceability and accountability in government dealings.

This feature is particularly beneficial in public procurement processes, where transparency has traditionally been challenging to achieve. Blockchain is a useful resource for governments to create trustworthy records resistant to manipulation which should serve for greater confidence in public institutions. The implementation of such technology holds the promise of minimizing fraudulent activities and ensuring that public resources are utilized effectively (Alotaibi et al., 2025).

Big Data Under the Trump Administration

In this already complex and constantly changing environment, the Trump administration's approach to big data and AI presents diverse opportunities and challenges. With significant changes to the U.S. government's handling of health data and other critical information resources, the administration has initiated sweeping reforms that prioritize deregulation and investment in AI infrastructure. For instance, a frenzy of activity surrounding major initiatives like Project Stargate highlights the administration's commitment to building AI data centers with substantial backing, including a proposed $500 billion investment from prominent tech companies (Sandler et al., 2025). This prioritization reflects a strategy to enhance the U.S. competitive edge in global AI development while navigating the increasing energy demands associated with data centers, which are anticipated to grow significantly in the coming years.

However, challenges remain as these regulatory changes raise concerns about transparency, accountability, and safeguarding public health information. The removal of extensive health-related data from federal websites has sparked criticism, as this information is vital to public decision-making and health management. Critics like Deborah Blum have expressed fears that these actions represent a significant contraction of public health knowledge and an erosion of trust in government (Giles, 2025).

Additionally, while the administration's regulatory rollbacks aim to stimulate innovation, they could inadvertently lead to risks of misinformation and abuse without sufficient oversight. Thus, businesses and stakeholders must adopt proactive strategies in compliance, partnerships, and energy procurement to effectively navigate this complex environment. Despite these concerns, the interplay between AI advancements and regulatory frameworks will continue to shape the future of technology and governance in the US.

Challenges in Maintaining Accountability

One major challenge in maintaining accountability stems from the natural resistance within government entities to transparency initiatives. This reluctance is often driven by a desire to avoid scrutiny and the potential ramifications that come with it. When governments resist efforts to be more open about their actions, it erodes public trust. Citizens feel disconnected from decision-making processes, leading to skepticism about whether the government acts in their best interests.

Another obstacle is the prevalence of information overload, which can inadvertently lead to disengagement rather than informed participation. In an age where data flows constantly, citizens may struggle to process and prioritize vast amounts of information. The sheer volume of data can be overwhelming, making it difficult for individuals to discern what is essential or trustworthy. As a result, excessive information sometimes leads to apathy instead of motivating active civic participation, as people withdraw from engaging with governmental issues due to mental fatigue.

Furthermore, the digital gap poses a significant barrier to government accountability and transparency. Disparities in technology access mean that not everyone can participate equally in digital transparency initiatives. Those without reliable Internet access or digital literacy skills are effectively excluded from the political process. This divide undermines the effectiveness of technology-driven governance strategies, as a portion of the population remains unrepresented in online discourse and decision-making. To bridge this gap, guidelines should be implemented to ensure equitable access to digital tools, allowing broader citizen participation across all socioeconomic backgrounds.

Trump's Resistance to Increasing Accountability: The Deep State

President Trump's second administration has introduced strategies that significantly limit accountability within the U.S. government, exemplified by the rapid rollback of transparency and oversight measures. Shortly after taking office, Trump initiated sweeping changes by purging critical health data from federal websites, including information pertaining to vaccines and other vital public health topics.

The administration's actions appear to align with a broader strategy to centralize power by dismissing officials deemed insufficiently loyal and undermining the checks and balances that ensure government accountability. In just a few weeks, sweeping alterations affected federal websites, such as those of the Centers for Disease Control and Prevention (CDC) and the Department of Health and Human Services (HHS).

The New York Times reported that more than 8,000 pages across a dozen government websites were removed in a span of two days. This purge eliminated critical information regarding vaccines, veterans' care, hate crimes, and scientific research, among other essential topics. Such actions can obstruct transparency and accountability by limiting public access to vital health information and undermining the ability of citizens to scrutinize and evaluate government actions effectively (Giles, 2025).

Moreover, Trump's approach includes leveraging the presidential authority to declassify information, which could potentially undermine national security and public trust. By emphasizing a lack of procedural constraints surrounding declassification, the administration risks transforming a tool meant to promote transparency into a mechanism for political gain. This manipulation of information both erodes public confidence and opens the door for selective disclosures that can mislead citizens and sway public opinion. The implications of these strategies extend beyond immediate governance; they threaten to normalize a culture of opacity and executive overreach that could persist across future administrations.

Ultimately, by prioritizing political expedience over accountability, Trump's actions reflect a substantial shift in the balance of power within the federal government, raising enduring concerns about the health of American democracy and the integrity of its institutions (Giles, 2025).

The War Against the Deep State

During Trump's second term, he has aggressively pursued his agenda to dismantle what he characterizes as the "deep state," targeting federal bureaucrats and agencies he deems insufficiently loyal. Since returning to the White House, Trump has targeted high-profile figures who criticized him. He revoked Secret Service protection for three former senior national security officials—John Bolton, Mike Pompeo, and Brian Hook—despite them being assassination targets on an Iranian government hit list (Tait, 2025). Trump has framed his actions as a crusade against what he refers to as "cancer" caused by the previous administration, initiating sweeping changes to government operations (McGreal, 2025):

> ***We're getting rid of all of the cancer – I call it cancer – the cancer caused by the Biden administration," he [Trump] told reporters as he signed an executive order stripping civil servants of many of their employment protections. (para. 11)***

Executive Order 14171, *Restoring Accountability to Policy-Influencing Positions Within the Federal Workforce*, includes stripping civil service protections from thousands of federal employees and encouraging mass resignations by offering lucrative incentives, such as eight months' pay for early retirement. Such tactics indicate a deliberate strategy to undermine the independence of the civil service and replace career officials with loyalists who align closely with Trump's political objectives (Oliphant & Holland, 2025).

The revocation of job protections and the purge of agency watchdogs—a move that has been criticized as an assault on accountability—reflect an intent to consolidate power within the executive branch (Oliphant & Holland, 2025). By removing oversight and entities that monitor government wrongdoing, Trump is potentially limiting scrutiny and creating an environment where federal workers may fear retribution for not adhering to his directives.

This trajectory has alarming consequences for governance, as it not only diminishes transparency but also threatens to establish a toxic atmosphere within federal agencies, undermining citizens' trust in their government. Critics have likened Trump's approach to authoritarian regimes, warning that such measures could pave the way for normalized political retaliation and the erosion of democratic accountability at the highest levels of government (Oliphant & Holland, 2025; Tait, 2025; McGreal, 2025).

Misinformation as a Political Resource

Trump's Strategy and Political Outcome

Trump's Executive Order 14179—*Removing Barriers to American Leadership in Artificial Intelligence*—has the potential to exacerbate the existing challenges posed by misinformation and falsehoods in the digital landscape. By effectively dismissing previous policies aimed at curbing misinformation—specifically those that targeted harmful narratives during the COVID-19 pandemic—Trump's order risks reinvigorating the spread of misleading information, thereby undermining public trust in governance and institutions.

The study of misinformation has evolved since 2016, revealing that prominent figures, including politicians, often disseminate harmful narratives, impacting public perception and political decisions. Experts note that the most significant misinformation often originates from influential domestic actors rather than anonymous online trolls (Clarke, 2025). This can lead to a disconnect in accountability and responsibility, especially when misinformation causes real-world harm, including threats against election officials during Trump's previous administration.

Additionally, Executive Order 14147, *Ending the Weaponization of the Federal Government,* may weaken the capacity of agencies like the CISA to effectively monitor and counter disinformation campaigns, particularly those originating from foreign adversaries. Critics argue that empowering misinformation rather than addressing it will embolden those who use such tactics to destabilize democratic processes (Swenson, 2025). Trump's focus on freedom of expression, while aiming to validate his supporters' viewpoints, may inadvertently contribute to the normalization of misinformation, complicating efforts to maintain information integrity in public discourse.

Trust Issues in Technology

As misinformation proliferates across digital platforms, public trust in both institutions and information systems is increasingly under strain. The manipulation of narratives through fake news not only distorts public opinion but influences policymaking itself. In a landscape shaped by rapid technological change, restoring trust has become an urgent imperative.

Technology companies sit at the center of the information ecosystem. Their platforms serve as primary conduits for news, making their role in curbing misinformation critical. Ethical standards must guide the flow of information, ensuring users receive accurate, verifiable data. Governments and tech firms alike must adopt proactive, trust-centered strategies—from algorithmic transparency to ethical data handling—to reinforce truthfulness in public discourse.

Building trust also requires engagement at the community level. Clear guidelines for how technologies interact with users—especially in civic and political contexts—can help rebuild fractured relationships between institutions and the public. Transparency isn't just technical—it's human.

One of the most powerful tools in digital trust-building is public code sharing. When government agencies open-source software developed with taxpayer funds, they invite public scrutiny, improvement, and collaboration. For example, the U.S. government's use of GitHub repositories—as seen with the Login.gov project—has enabled outside experts to audit systems, discover vulnerabilities, and improve user experience.

These community contributions led to measurable changes: updated encryption protocols, clearer documentation, and fewer authentication errors. Unlike proprietary software, which limits accountability to internal vendors, open code offers transparency, resilience, and shared ownership.

This spirit of openness is also formalized through bug bounty programs, where ethical hackers are rewarded for identifying software flaws. These systems help catch problems before they escalate, strengthening public confidence in digital governance. Open discovery processes reduce hidden backdoors, foster continuous improvement, and create shared accountability—a sharp contrast to closed-door software ecosystems.

As agencies adopt AI to sort applications or flag fraud, the demand for explainability becomes essential. If a person is denied access to a service by an algorithm, they should know why. Tools developed for the Department of Labor's modernization effort now offer not only outcomes, but also transparent explanations of the inputs and logic behind decisions.

This push toward explainability has already yielded positive results. By translating complex machine learning outputs into plain language—often in partnership with user research teams—agencies like the CDC have improved public understanding and reduced complaints. When people understand how a system works, trust increases, and resistance declines.

The Freedom of Information Act (FOIA) was designed for paper records—but today, information lives in encrypted chats, cloud docs, and apps. Modern FOIA tools now include integrated archival systems that capture, tag, and store digital communications across multiple formats. The Department of Homeland Security, for instance, has implemented searchable encrypted logs and file storage to keep up with the evolving nature of records.

New AI systems help categorize and redact documents faster, reducing bottlenecks and improving response times. Where FOIA requests once lingered for months, automation and better archiving have led to shorter wait times and more equitable public access.

Transparency has also moved into real-time oversight. Dashboards displaying project milestones, service uptimes, and government spending are now visible to the public. Initiatives like the General Services Administration's login modernization dashboard allow anyone—journalists, watchdogs, or citizens—to monitor performance and hold agencies accountable.

The impact is already visible: before dashboards, quarterly reports were often delayed and vague. Now, users can track system status as it happens, reducing speculation and increasing public confidence in agency responsiveness.

Transparency initiatives—such as open-source code, explainable AI, upgraded FOIA systems, and real-time dashboards—are not just technical upgrades. They represent a cultural shift in how the public expects government to operate: with accountability, openness, and inclusion. These tools help address the deeper governance challenges discussed throughout this chapter. They also raise new questions about bias, exclusion, and dependency on private vendors—topics that remain central in any conversation about ethical digital governance.

Alongside institutional reforms, public education remains essential. Promoting media literacy and supporting fact-checking initiatives can reduce the societal impact of misinformation. Collaborating with educators helps equip individuals to navigate digital landscapes with greater discernment—ensuring that innovation is matched by an informed and resilient society.

Societal Implications of AI-Driven Governance

The growing presence of artificial intelligence in government isn't just a technological shift—it's a transformation in how power operates. When algorithms shape public policy, the stakes rise. These systems aren't neutral; they inherit the biases of their creators and the datasets they're trained on. Without proper oversight, AI can quietly reinforce inequality while giving the illusion of objectivity.

That's why transparency matters. Citizens deserve to know how decisions are made—especially when they're made by machines. And yet, marginalized communities are often the first to feel the consequences of flawed algorithms. If we're serious about fairness, we must ask: Who is building these systems? Whose values are embedded in the code? And who gets to opt out?

To protect the public interest, algorithmic governance must not be allowed to outpace democratic safeguards. The call isn't just for better technology—it's for better accountability.

Technology has opened new doors for civic engagement. Platforms like Twitter, Instagram, and even government-run portals are replacing town halls and phone banks. They offer real-time feedback loops where citizens can speak directly to those in power—and just as importantly, to each other.

These digital forums democratize access to dialogue. They allow movements to form quickly, voices to be amplified, and policies to be challenged publicly. When designed with care, they can make democracy more accessible—not just louder.

But access doesn't always equal influence. We still need to ask: Who is being heard? Who is being silenced by algorithmic prioritization or harassment? For these platforms to truly serve the public, they must be as committed to equity as they are to engagement.

The old ways of gathering public opinion—phone calls, door-to-door surveys—are fading fast. Caller ID, spam filters, and shifting communication habits have pushed pollsters online. But even there, challenges remain. Internet panels can be skewed, leaving out the voices of those without access or interest.

What we see today is a polling landscape in flux. Some polls are backed by campaigns with selective transparency, strategically releasing favorable results. Others rely on methodologies that fail to reflect the complexity of public sentiment.

There's a growing push for more rigorous, nonpartisan polling—often from academic institutions or transparent research centers. Still, even the best surveys must reckon with distrust and a fragmented media environment.

Restoring faith in polling requires more than better tools—it calls for honesty about their limitations and a commitment to capturing nuance, not just numbers.

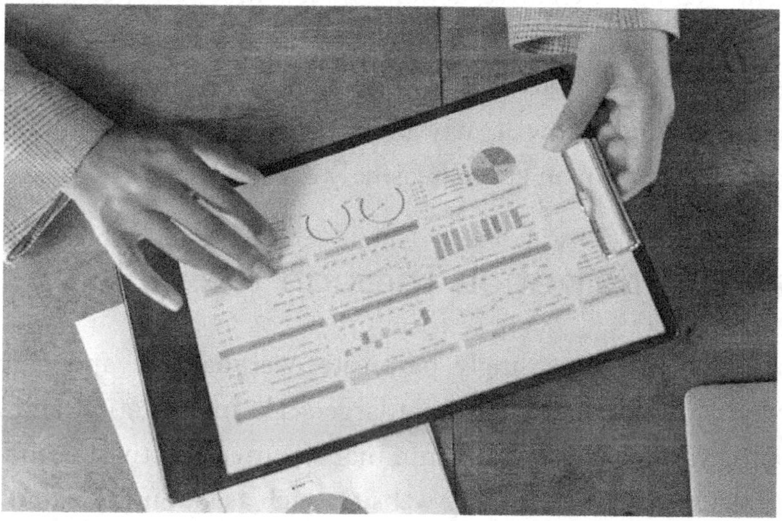

The transition to Trump's administration has elicited mixed perspectives from citizens regarding the role of AI and technology. While there is optimism surrounding the potential for AI to drive innovation and establish the US as a leader in the tech space, concerns about its impact on accountability, transparency, and employment remain prevalent. A significant portion of the population views AI as a double-edged sword; a Pew Research Center survey indicates that many Americans believe advancements in AI could lead to the creation and distribution of misleading information, contributing to fears of manipulation during critical events such as elections (Gracia, 2024).

This skepticism is further compounded by the perception that technology companies while benefiting from deregulated environments, may not effectively safeguard their platforms against misuse. Despite a majority of Americans advocating for tech companies to take responsibility for preventing harmful outcomes, confidence in their ability to do so has sharply declined, with only 20% expressing assurance that these firms will successfully manage risks associated with their technologies (Gracia, 2024).

President Trump's executive actions in his first few weeks in office showcase the uncertainty that lies ahead, with many diverse voices emphasizing the necessity for public education on technology. Concerned specialists highlight how it may impact ensuring understanding and responsible growth in an era increasingly dominated by AI.

Citizen Participation Through Technology

Digital platforms increasingly permeate everyday life and empower grassroots movements by enabling large-scale mobilization and fundraising for various causes with minimal resources. Activists can utilize social media and crowdfunding websites such as GoFundMe and Kickstarter to amplify their voices and secure financial support for initiatives that may otherwise struggle for backing. Successful campaigns in environmental and social justice spheres illustrate how these tools can sustain momentum and achieve objectives.

Moreover, social media platforms serve as vital spaces for marginalized voices, allowing underrepresented communities to share their stories and experiences. Movements like #MeToo demonstrate the power of digital engagement in spotlighting societal issues, resulting in significant policy changes. Technology also supports democratic processes by gathering diverse opinions on policy issues through digital surveys and interactive forums, enhancing representative decision-making.

Online consultation platforms allow governments to solicit feedback on proposed policies, ensuring various perspectives are considered for more equitable outcomes. Digital petitions, facilitated by platforms like Change.org, also mobilize public support for reforms, pressuring lawmakers to act on pressing societal issues. Overall, digital advocacy campaigns enrich political discourse, empowering citizens to express their concerns, hold officials accountable, and participate actively in shaping their governance. Virtual town halls further enhance transparency, enabling inclusive dialogue between citizens and their representatives, ultimately leading to improved accountability and governance.

In times when governments may resort to censorship and heightened street control to suppress dissent, technology emerges as a crucial tool for individuals to express their political opposition. Digital platforms, social media, and encrypted communication channels provide activists and marginalized voices an avenue to share their perspectives, organize protests, and mobilize support without relying on traditional and potentially censored means of communication.

These technological tools enable citizens to bypass restrictions imposed by authoritarian regimes, instilling a sense of agency and solidarity among those advocating for change. As governments intensify their efforts to stifle free expression, the strategic use of technology can play an essential role in sustaining movements for political dissent and ensuring that diverse viewpoints remain visible and heard.

Technology and Political Expression of Dissent

During Trump's second term, his administration has employed a multifaceted strategy to control protests and influence public opinion through technological means. Amid rising protests in response to various policies—including immigration crackdowns and the rollback of individual rights—Trump has sought to delegitimize dissenting voices by framing them as part of a broader narrative against his administration.

In an effort to bolster free speech by signing Executive Order 14179, *Removing Barriers to American Leadership in Artificial Intelligence*, Trump ostensibly reaffirmed First Amendment rights. Yet, the order paradoxically serves as a means to stifle criticism by accusing previous administrations of censoring conservative voices.

Additionally, Trump's push to diminish civil rights protections further contributes to a culture of surveillance and control as he seeks to expand mass surveillance capabilities. Trump's Executive Order 14147, *Ending the Weaponization of the Federal Government*, places significant power in the hands of federal officials, allowing them to investigate and prosecute individuals with considerable discretion. This power can lead to devastating consequences for targeted individuals, as it opens the door for potential abuse, such as using investigations and prosecutions to target political opponents or discriminate against marginalized communities. Trump's threats to leverage agencies like the DOJ for political gain and demand loyalty from civil servants further exacerbate concerns about the erosion of legal norms and the integrity of civil rights protections (*Trump on Surveillance, Protest, and Free Speech*, n.d.). This climate raises concerns about the consequences for individual rights and freedoms, particularly among marginalized communities. As public protests grow in response to governmental actions, the administration's approach highlights the tension between preserving civil liberties and enforcing a narrative that seeks to consolidate power.

A Blueprint for Transparency and Accountability

In response to these trends, advocates have proposed a bold set of reforms aimed at making AI governance more transparent, participatory, and just:

1. **Real-Time AI Dashboards**
   Require federal agencies to display ongoing AI use and policy shifts in public dashboards, open to scrutiny and comment.

2. **Modernized FOIA**
   Expand the Freedom of Information Act with a digital portal that auto-logs requests and tracks compliance.

3. **Independent Oversight Commission**
   Establish a watchdog group of technologists, ethicists, and citizens to audit government use of big data and AI.

4. **Open Code and Clear Summaries**
   Mandate that all AI systems include open-source code and plain-language summaries explaining their public impact.

5. **Annual AI Ethics Hearings**
   Hold Congress-led public hearings where federal leaders must account for algorithmic decisions and civil liberties safeguards.

6. **Regulatory Impact Reviews**
   Require evaluations of all proposed rollbacks in tech-related regulations, measuring transparency and social consequences.

7. **Whistleblower Protection**
   Bolster protections and fast-track complaints tied to data misuse or unethical surveillance.

8. **Contractor Accountability**
   Tie future contracts to transparent audits of past work, encouraging responsible innovation.

9. **Civic-Tech Fellowships**
   Embed developers and journalists within agencies to create tools that translate policy into public understanding.

10. **Sunset Clauses**
    Add expiration dates to deregulation measures, forcing periodic review based on real-world data and feedback.

These aren't just technical fixes, they're a commitment to keep governance human-centered in an increasingly automated world.

The Pushback: What Could Go Wrong?

Of course, no reform is without risk. Critics from across the political spectrum raise concerns, including:

1. **Overregulation Risks**:
   Critics may argue these measures could overburden agencies with compliance tasks, delaying innovation and responsiveness in fast-moving tech environments.

2. **Chilling Innovation**:
   Mandatory audits, algorithmic disclosure, and public impact summaries might deter private sector partnerships with government agencies, limiting access to cutting-edge technologies.

3. **Security and Privacy Concerns**:
   Publishing source code and data logs could expose sensitive infrastructure details, raising cybersecurity risks or compromising national security protocols.

4. **Politicization of Oversight**:
   Independent commissions and ethics hearings may become battlegrounds for partisan grandstanding, diminishing constructive discourse and policy consistency.

5. **Resource Intensity**:
   aunching civic-tech fellowships, ombuds offices, and centralized portals could be costly and difficult to sustain without guaranteed funding or cross-agency support.

6. **Reduced Agility**:
   Sunset provisions and mandatory impact assessments could slow policy adaptation, hindering the government's ability to respond nimbly to emerging challenges.

7. **Contractor Reticence**:
   Tying procurement eligibility to public audits may discourage top-tier vendors from engaging in federal innovation projects due to reputational risk.

8. **Complexity for Citizens**:
   Lay-friendly summaries and dashboards may still be too technical or vast for average citizens to meaningfully engage with, creating a veneer of accessibility without true empowerment.

9. **Duplication and Bureaucracy**:
   A new layer of oversight entities and review mechanisms might duplicate existing efforts, creating redundancy rather than efficiency.

10. **Risk of Entrenching Elites**:
    Embedding journalists and developers in agencies could concentrate influence among a select group, unintentionally sidelining broader public input or accountability.

These warnings highlight the balancing act ahead. Accountability must not come at the cost of agility. But nor should speed override justice.

As AI-driven governance takes hold, one truth remains: the public deserves a seat at the table—before the decisions are made, not just after they're felt.

Bringing it All Together

- **AI innovation is accelerating** under a deregulatory agenda, potentially boosting U.S. leadership in the global tech economy.
- **Accountability and transparency are at risk** as public-facing data becomes harder to access and oversight mechanisms are weakened.
- **Public health and safety may suffer** when crucial data and infrastructure fall outside the bounds of transparency.
- **Citizen awareness is diminishing** as the gap widens between government actions and public understanding.
- **The overall trend points toward centralized control**, where rapid innovation may come at the cost of democratic safeguards.

Chapter 7

Democracy in the Era of Technology

The impact of technology on democracy has become increasingly pronounced, especially in the wake of economic challenges faced globally in 2024. The pervasive use of digital platforms and social media has transformed how citizens engage with political processes, but it has also highlighted significant discontent with representative governance. Surveys indicate that a median of 54% of adults in 31 countries expressed dissatisfaction with democracy, revealing a widespread perception that political leaders and institutions are disconnected from the public. This sentiment is exacerbated by the rapid dissemination of misinformation, which threatens the integrity of information and public discourse essential to democratic functioning (Wike et al., 2024).

As populist movements, such as Trump's "Make America Great Again," gain traction, existing governmental structures are being tested. The technological realm demands that democracies bolster transparency and accountability and combat the spread of false information and foreign interference. Emphasizing the need for citizen trust, the OECD Trust

Survey illustrates how governments must actively demonstrate their commitment to addressing citizens' interests and involve them in complex decision-making processes (*2024 Global Forum on Building Trust*, 2024). Thus, collaborative efforts between governments, civil society, and the private sector—both nationally and internationally—are crucial.

The interconnected challenges of the 21st century require democratic systems to reinforce their resilience through cooperation, fostering an environment where public trust can flourish and ensuring that citizen voices are respected and integrated into the political level. As technology evolves, it intertwines with every aspect of life, shaping societies in ways we are only beginning to comprehend. This is especially evident in the balance between maintaining national security and safeguarding individual freedoms—a delicate act that modern democracies must navigate carefully.

With sophisticated surveillance technologies becoming more embedded in daily operations, a pressing conversation emerges around their potential impact on civil liberties. How these tools might infringe upon personal freedoms and the ethical dilemmas they pose could redefine democratic principles if not addressed responsibly. Let's now discuss how technological advances intersect with democratic ideals.

Risks of Autocratic Tendencies

As governments and corporations adopt increasingly advanced surveillance tools, the balance between national security and civil liberties grows more fragile. Without proper oversight, these technologies risk infringing on individual rights—often silently and without recourse. This evolving reality raises critical ethical questions about how much power should be ceded to systems that can operate beyond the public's view.

At the same time, the weaponization of information through digital means poses a growing threat to democratic institutions. Advances in generative AI have made it easier than ever to fabricate convincing narratives, blurring the line between fact and fiction. Disinformation campaigns, once difficult to orchestrate, now scale effortlessly—amplifying bias, eroding public trust, and sowing confusion. Studies show that AI-powered chatbots, including ChatGPT and others, have at times unknowingly circulated misleading content, unable to consistently identify or flag unreliable sources (Csernatoni, 2024). The challenge ahead isn't just about regulating content—it's about preserving truth itself.

The Disinformation Playbook

Have you ever paused to consider that many of the posts you scroll past might not be casual opinions—but deliberate tactics? Behind what often feels like harmless noise lies a strategy to confuse, manipulate, and reshape public belief. This is the new disinformation playbook: bots, deepfakes, clickbait, and algorithm-driven echo chambers working in concert to distort perception.

Bot Networks

Automated bots now mimic grassroots movements, flooding social media with coordinated content that feels organic but isn't. During a simulated election, over 5,000 bots used hashtags like #ElectionRigged and repeated emotional trigger words like "fraud" and "betrayal" during peak hours (Samson Olufemi Olanipekun, 2025). Their goal? To saturate timelines, hijack visibility algorithms, and create the illusion of consensus—ultimately manufacturing distrust and division.

Deepfakes

Using AI to generate near-perfect video and audio, deepfakes manipulate what we see and hear. In one viral case, a video showed a respected leader endorsing a controversial stance—only it never happened. The video racked up over a million views before being debunked (Olanipekun, 2025). Even after exposure, the damage lingered. Viewers were left questioning reality, and the emotional residue of doubt remained long after the truth was known.

Clickbait Economics

Misinformation isn't just powerful—it's profitable. Sensational headlines and outrageous claims generate clicks, which drive ad revenue. During the 2016 election, one fake news article earned $5,000 per hour at its peak (Wired, 2024). Platforms are rewarded for engagement, not accuracy, creating a business model that favors speed, shock, and virality over fact-checking or journalistic integrity.

Echo Chambers

Social media algorithms feed users more of what they already believe, reinforcing biases and creating silos of thought. Within these bubbles, false information thrives, bolstered by repetition and peer validation. Efforts to correct misinformation often fail—not because the facts aren't clear, but because they're unwelcome. The result is a population increasingly divided, less willing to consider alternative views, and more vulnerable to radicalization.

Together, these tactics form an ecosystem of digital manipulation. Bots drive visibility, deepfakes distort perception, monetized misinformation fuels the fire, and echo chambers lock it in. The combined effect isn't just confusion—it's corrosion. Public trust suffers, civil discourse erodes, and the line between propaganda and reality becomes dangerously thin.

Even as disinformation grows more sophisticated, government oversight tools remain outdated. Traditional filters and static monitoring methods are no longer enough. Combatting this new era of information warfare demands agile systems, real-time data analysis, and a reimagined approach to digital accountability—one that can adapt as quickly as the threats themselves evolve. Without it, democracy risks being reshaped by those who master the manipulation first.

Concentration of Uniform Political Conversations

The centralization of power within tech monopolies further complicates the political landscape. Major technology firms possess immense influence over public discourse due to their control over social media platforms, data analytics, and content moderation (Shahbaz, 2018). This monopolistic control stifles competition and presents a risk to diverse political conversations. In some instances, algorithms on popular platforms like TikTok have reportedly amplified content favoring certain political narratives, underscoring concerns about algorithmic transparency and its impact on political outcomes.

In the age of digital communication, political censorship remains another contentious issue. The pressure to implement censorship measures often stems from fears of destabilizing misinformation or foreign interference but risks evolving into a tool for suppressing legitimate dissent. Platforms wield significant authority in determining what constitutes permissible content, which can lead to the suppression of dissenting voices and the restriction of open political speech. Thus, governments around the world, including democratic nations, are grappling with defining and implementing rules that strike a balance between preventing harmful content and upholding freedom of expression.

Recognizing the potential threats posed by the centralization of tech power, it's crucial to consider strategies for maintaining balanced governance. Guidelines that encourage alternative, decentralized Internet governance can diminish the overpowering influence of technology corporations on public discussion. Promoting open-source software, nurturing competitive marketplaces, and backing community-driven projects can enhance diversity and robustness within the digital context.

Furthermore, partnerships among civil society, governmental bodies, and private enterprises are crucial for crafting regulations that preserve democratic principles while adapting to technological progress. For example, the partnerships in Italy and the US aimed at increasing media literacy highlight the benefits of collaborative efforts in combating the spread of misinformation (Shahbaz, 2018). On April 17, 2024, the U.S. Secretary of State and Italy's Vice President of the Council of Ministers signed a memorandum of understanding (MOU) geared toward enhancing collaboration between the US and Italy in combating foreign state information manipulation. This manipulation includes disinformation and propaganda, which pose a transnational security threat, creating or exploiting divisions within and among nations, jeopardizing election integrity, and undermining public trust in government (*The United States of America and Italy*, 2024).

Trump's Policies and a Potential Impact on the United States Democracy

Under President Trump's renewed administration, his explicit mission to dismantle what he terms the "deep state" reflects a deliberate strategy to centralize executive power and reduce independent oversight across the federal government. While this theme has been addressed earlier in the chapter, it bears repeating in this context: Trump's policies represent not just political rhetoric but a structural shift in how power is wielded within American democracy.

One of the most consequential moves was the signing of Executive Order 14147, *Ending the Weaponization of the Federal Government*, which strips away longstanding protections for civil servants. This order paves the way for the broad removal or replacement of experienced federal employees with individuals loyal to the administration—effectively converting nonpartisan institutions into partisan extensions of executive will.

Beyond this order, Trump's administration also orchestrated the abrupt dismissal of several inspectors general—independent officials responsible for internal accountability—without providing the legally mandated notice to Congress. These watchdogs, essential to preserving transparency and enforcing ethical standards, were removed in ways that undermined the very checks and balances foundational to democratic governance (Katz & Urby, 2025).

Critics, including legal experts and former officials, warn that this trajectory could normalize executive overreach and disrupt how the government interacts with citizens (Stone, 2025). The perception that the administration seeks to operate with minimal accountability under the guise of efficiency presents a significant threat to the foundational norms and values of U.S. democracy and subsequently, how it impacts the global network.. Ultimately, while Trump frames these actions as necessary for revitalizing governmental function, they pose a risk of damaging the very fabric of democratic accountability and oversight that is crucial for protecting citizens' rights and interests (Stone, 2025).

Surveillance Technologies and Privacy

In today's digital age, the intersection of democracy and technology brings to light profound implications for personal privacy and individual freedoms. One of the most contentious issues in this domain is the widespread use of surveillance technologies by governments and corporations. These practices not only intrude on personal privacy but also challenge democratic societies, raising questions about the balance between security and personal freedom.

Government surveillance programs, often shrouded in secrecy, pose a significant threat to individual privacy. These programs operate with minimal transparency, leveraging extensive data collection that extends beyond what many citizens are aware of. This can erode trust in governmental authority as individuals become increasingly skeptical of how their data is used or potentially abused. The lack of awareness among citizens regarding surveillance practices nurtures an environment where unchecked power can flourish, undermining the democratic principle of accountability (U.S. Government Accountability Office, n.d.).

Beyond government entities, corporations also engage in pervasive data-collection practices that similarly challenge individual autonomy. Companies collect vast amounts of personal information, often without explicit consent from users, raising critical concerns about data ownership and personal rights and leaving consumers feeling powerless over how their data is utilized. In light of these practices, questions about consent and data protection become ever more pressing. Users may find themselves unwittingly sharing sensitive information, which can be used for purposes ranging from targeted advertising to more insidious forms of manipulation (*Government Surveillance*, n.d.).

Ongoing Legislation

Existing privacy legislation struggles to keep pace with rapid technological advancements. Many laws, such as the Privacy Act of 1974 and the E-Government Act of 2002, were established before the digital revolution and, therefore, lack the robustness needed to protect individual freedoms effectively today. Without comprehensive federal privacy laws governing both the public and private sectors, there are gaps that leave personal data vulnerable. As technology continues to evolve, advocacy for stronger protections must intensify to ensure legislative frameworks are updated to reflect current needs and challenges.

Given these challenges, it also becomes vital to establish guidelines for corporate data collection and privacy legislation. Corporations should be transparent about their data practices, providing clear information about what data is collected, how it is used, and whom it is shared with.

Policies that require explicit user consent before data collection and allow users to access, amend, or delete their data will build the steps toward empowering users and protecting privacy.

Similarly, privacy legislation needs to be proactive rather than reactive. Laws should anticipate future technological developments and adaptively incorporate measures that safeguard personal freedoms. Such legislation could include mandatory impact assessments for new technologies, regular audits of data handling practices, and stringent penalties for noncompliance. Policymakers should prioritize creating comprehensive federal privacy laws that encompass all aspects of data collection and surveillance, thus closing existing loopholes (U.S. Government Accountability Office, n.d.).

Trump's Surveillance Program

During President Trump's second term, there are heightened concerns regarding the use of surveillance strategies as tools of power, particularly against political enemies and vulnerable populations. Experts warn that Trump may exploit U.S. intelligence agencies to conduct extensive mass surveillance, leading to a significant erosion of individual rights (Friedman & Delerme, 2025). Mass surveillance typically involves the widespread monitoring of civilians, with governments often targeting specific groups based on race, religion, or immigration status. Such practices can facilitate measures to "contain" these populations, including arrests and imprisonment, posing serious threats to civil liberties.

The broad surveillance capabilities, bolstered by programs like Section 702 of the Foreign Intelligence Surveillance Act (FISA) (2008) and Executive Order 12333 (1981), have allowed the government to collect and access vast amounts of personal data. These tools enable dragnet surveillance without appropriate warrants or safeguards and have raised concerns about privacy violations since their enactment. With Trump's past intentions to target political opponents, there is a legitimate fear that these surveillance powers could be wielded to discriminatorily monitor and harass dissenters, including protesters, people of color, immigrants, and individuals seeking abortion or gender-affirming care. The potential abuse of executive authority in this context places immense power in the hands of federal officials, granting them the ability to surveil Americans' lives with little oversight.

The implications of such surveillance strategies extend beyond individual privacy; they reverberate through the very fabric of democratic society. As civil liberties are eroded, public trust in government institutions falters, creating a chilling effect on free speech, dissent, and civic participation. Organizations like the American Civil Liberties Union (ACLU) have begun closely tracking these developments, signaling potential legal challenges and advocacy campaigns to safeguard constitutional protections. Their involvement underscores the urgent need for robust checks and transparent oversight to prevent executive overreach.

What is at stake is not simply a matter of policy—it is the trajectory of democracy itself. If left unchecked, the Trump administration's expanding surveillance agenda could normalize an authoritarian model of governance, one where control is exercised not through consent but through coercion, and where surveillance is used not as a shield, but as a sword (Friedman & Delerme, 2025).

Big Data, Algorithm Biases, and Censorship

The influence of big data analytics on personal autonomy and decision-making is significant in today's technology-driven world. While insights derived from vast datasets can enhance efficiencies across sectors, they also raise privacy and individual freedom concerns.

Data profiling, which involves creating detailed profiles to predict behaviors and preferences, poses risks of discrimination and inequality, particularly if algorithms reflect biases that disadvantage underrepresented groups (Howe & Elenberg, 2020). Moreover, data-driven algorithms can manipulate consumer choices subtly through tailored content, challenging the authenticity of personal autonomy. This manipulation can influence political beliefs and consumer habits, rendering individuals' freedom to choose an illusion (Lu, 2024).

The ethical implications necessitate informed consent, which empowers users to understand how their data is utilized, yet maintaining transparency in the rapidly evolving tech environment proves difficult. Additionally, big data's impact extends to personal relationships via social media, where algorithms can create echo chambers that reinforce biases and diminish meaningful interactions. To address these challenges, technologists and policymakers advocate for improved algorithm transparency, responsible AI development, and regulations to safeguard individual rights while enhancing user agency over personal data (Lu, 2024).

Trump - Freedom of Speech Versus Respect

Upon returning to the White House, Trump faced significant protests against his administration's policies on immigration, civil rights, and health. In response, he has sought to delegitimize dissent by framing protesters as threats to national security and portraying dissenting views as part of a broader conspiracy against his presidency. For instance, Trump has framed the situation at the U.S.-Mexico border as an "invasion" and a national emergency. This rhetoric sets a dramatic tone for his administration, leading to actions such as deploying thousands of troops, invoking the Alien Enemies Act (1798), and shutting down humanitarian programs (Chishti & Bush-Joseph, 2025). By leveraging social media platforms, Trump amplifies these narratives, reaching a vast audience to rally support while discouraging opposition.

Trump's Executive Order 14149, *Restoring Freedom of Speech and Ending Federal Censorship*, promises to secure the right to free speech and correct federal censorship in online forums (The White House, 2025b, sect. 2). While the intention to bolster free speech may resonate positively with many, a deeper examination reveals that this order could ultimately undermine the very democratic principles it purports to uphold.

Central to this concern is the premise upon which the order is based, which inaccurately claims that the Biden administration "trampled free speech rights" (The White House, 2025b, sect. 1) in its efforts to promote what it deemed a responsible public narrative regarding issues like the COVID-19 pandemic. In reality, the Biden administration aimed to safeguard public health by curbing the spread of dangerous disinformation—specifically misinformation regarding vaccines, which had tangible life-or-death repercussions. This disingenuous framing not only threatens to distort the public understanding of free speech but also underscores the potential for eroded trust in institutions that aim to protect the populace from misinformation, ultimately compromising the integrity of the democratic process.

Meta and X: Trump's Search for Allies

As part of this strategy, Trump has dismantled the established fact-checking mechanisms on platforms like Meta on Facebook and Instagram. Meta's fact-checking program, implemented on Facebook and Instagram in 2016, aimed to combat the spread of false information by referring posts that appeared misleading to independent organizations for verification.

This system allowed flagged posts to receive labels providing additional context, often reducing their visibility in users' feeds as a means of discouraging misinformation. However, under the recent policy shift initiated by Trump, this fact-checking system is being replaced with a model akin to X's "community notes," which relies on user-generated content to add context or clarity to controversial posts. This change is seen as a move away from rigorous fact-checking toward a system that may facilitate the spread of misinformation, echoing the broader abandonment of accountability seen in Trump's other policies.

The decision to eliminate third-party fact-checking is troubling, especially given concerns that this could lead to a rise in unchecked harmful content and hate speech on social media platforms. Critics argue that Meta's abandonment of independent moderators signals a capitulation to political pressures, especially with Trump's overt attempts to influence social media narratives and diminish oversight mechanisms (McMahon et al., 2025).

This can be considered a sign amid a broader trend in which the governance of online speech is increasingly politicized, raising alarms about the implications for public discourse, the integrity of information, and the erosion of safeguards against harmful content, ultimately blurring the lines between freedom of expression and accountability in the digital space. This step away from independent verification of content allows misinformation to thrive, which is justified by claims of a need for "free expression.

By portraying his actions as efforts to combat censorship, Trump can consolidate political support among his base while undermining accountability. Furthermore, by labeling opponents and critics as representatives of a corrupt establishment, Trump rallies his supporters as he creates an environment where dissent can be met with surveillance and intimidation.

Balancing Security and Individual Rights in a Tech-Driven World

The Role of States

Government surveillance, while often justified as a tool to combat threats such as terrorism or cybercrime, must be carefully assessed to determine its proportionality (Königs, 2022). In democratic nations, surveillance is typically framed as a means of maintaining public safety and order. However, the potential for overreach remains a constant concern. Citizens fear that pervasive government surveillance could lead to an erosion of privacy, where personal data is collected en masse without adequate checks and balances (*Guiding Principles on Government Use of Surveillance Technologies*, 2023).

Public accountability mechanisms play a pivotal role in preventing security measures from overreaching. These contribute to a transparent oversight of processes to monitor and evaluate government actions, ensuring that surveillance practices do not infringe upon civil liberties. Such mechanisms promote trust between the government and its citizens. Thus, people feel more secure knowing there are systems in place to check power and prevent abuse (*Guiding Principles on Government Use of Surveillance Technologies*, 2023).

Companies

The advancement of technology requires companies to balance user rights with innovation. This means acknowledging and taking action on how technological progress affects privacy and civil liberties right from the design and implementation stages. To achieve these goals, companies ought to embed ethical principles in their product development processes.

Companies—particularly large tech giants—are predominantly driven by profit motives and their own interests, often prioritizing shareholder returns over the common good. This profit-centric approach can lead to practices that undermine individual rights and societal well-being, as these corporations may exploit personal data or foster environments that enable the spread of misinformation.

Consumers' Responsibilities

Yet, ethical considerations aren't limited to companies; they also include consumer actions. As consumers gain a deeper understanding of privacy concerns, they can make educated choices regarding the products and services they opt for. Grasping the ethical dimension empowers consumers to favor companies that share their values, thus driving demand for ethical technology solutions and shaping industry norms.

On the other hand, citizens should demand transparency in how their data is collected, stored, and utilized, advocating for clear communication regarding surveillance practices. Governments, in turn, must ensure that any data collected is handled securely and used only for its intended purpose to prevent misuse and safeguard individual privacy (*Guiding Principles on Government Use of Surveillance Technologies*, 2023).

Democratic Systems

In modern democracies, the state ultimately influences the balance between user rights and innovation, which is essential for safeguarding the integrity of the system. This involves actively safeguarding user rights through laws and policies that create a framework to establish transparency and accountability from tech companies.

In contrast to companies, a democratic government prioritizes the common good and is compelled to implement regulations and ethical standards that govern how technology is developed and utilized. Ethical behavior in this context means ensuring the responsible use of technological tools to protect people's safety, promote transparency, and maintain institutional stability rather than concentrating power in the executive branch as reflected in Trump's strategies.

A focus on the welfare of citizens and addressing the challenges posed by rapidly advancing technologies implies that the government creates a regulatory framework that encourages innovation while preserving democratic values. This approach might nurture a healthy democracy, where technological advancements benefit society as a whole rather than serve the narrow objectives of a few powerful entities.

Individual Strategies for Fighting Misinformation

While these strategies are geared toward individual action, they're just as valuable for community groups and local organizations committed to truth and media literacy.

1. **Verify the Source**: Before sharing or believing a claim, go to the original source. Look for an author, publication date, and citations to reliable data or institutions.

2. **Cross-Check Multiple Outlets**: Search for the same story across several reputable news organizations. If it's only showing up on obscure or highly partisan sites, approach with skepticism.

3. **Evaluate the Author's Credentials**: Look up the author's background. Experts typically have a history of credible work, academic ties, or recognized experience in the field.

4. **Inspect the Evidence**: When possible, locate primary sources—such as original studies, official reports, or transcripts. Review the methodology to ensure conclusions align with the data.

5. **Check for Expert Consensus**: Look for commentary or statements from respected experts or organizations. When there's widespread agreement among professionals, the information is more likely to be reliable.

6. **Use Trusted Fact-Checking Sites**. Save tools like Snopes, FactCheck.org, and PolitiFact. Search for the claim or headline to see if it's already been reviewed.

7. **Watch for Emotionally Charged Language**: Sensational headlines that trigger outrage or fear often rely on exaggeration. Take a breath before reacting or reposting.

8. **Examine the URL and Website Design**: Check for suspicious URLs (e.g., ".com.co" or odd spellings) and sloppy layouts. Legitimate outlets tend to use secure protocols (HTTPS) and professional formatting.

9. **Pause Before You Share**: In fast-moving news cycles, misinformation spreads quickly. Read beyond the headline and take a moment to assess before hitting "share."

10. **Talk It Out**: Discuss questionable content with someone you trust—especially if they bring a different perspective. A thoughtful conversation can reveal blind spots or red flags.

Community-Based Strategies for Fighting Misinformation

1. **Create Fact-Checking Hubs:** Host workshops at schools or libraries to teach source verification.

2. **Run Media Literacy Campaigns:** Distribute simple guides on spotting bias and algorithmic influence.

3. **Form Watchdog Groups:** Monitor local platforms and respond to false claims quickly.

4. **Host Public Forums:** Invite experts and locals to open Q&A sessions on key issues.

5. **Collaborate with Local Media:** Add "Myth vs. Fact" segments in local news and radio.

6. **Build Shared Resource Libraries:** Provide community access to trusted data and fact-checking tools.

7. **Train Community Ambassadors:** Equip local leaders with skills to counter misinformation.

8. **Deploy Rapid Response Teams:** Issue timely corrections on social platforms when falsehoods spread.

9. **Offer Digital Literacy Classes:** Teach practical tech skills to spot fake content and deepfakes.

10. **Track Progress and Adjust:** Survey community understanding and improve efforts as needed.

Bringing it All Together

- Surveillance under Trump raises civil liberty concerns.
- Targeted monitoring affects marginalized groups most.
- Executive power is being expanded at the cost of oversight.
- Dissent is increasingly framed as a security threat.
- Rollback of civil protections weakens democratic safeguards.
- AI in security poses ethical risks and power shifts.

Chapter 8

AI, Surveillance, and Military Norms

Imagine a near future where autonomous drones powered by artificial intelligence patrol international borders, scanning vast landscapes without pause. Meanwhile, cyberattacks grow more frequent and sophisticated, intensifying tensions between global powers. In cities, citizens navigate a delicate balance between security and privacy, as surveillance systems grow more pervasive.

This isn't science fiction—it's the emerging reality reshaping military operations, national security, and civil liberties.

Technology on the Frontlines

Technological advancements in military AI, cybersecurity, and surveillance are transforming how governments defend themselves and engage with the world. These tools bring undeniable benefits—real-time data analysis, faster decision-making, and enhanced strategic precision.

Yet they also spark urgent ethical debates. As machines assume roles once reserved for human commanders, including making life-or-death decisions, questions about accountability, transparency, and democratic oversight come to the forefront.

The Trump Administration's Strategic AI Push

Under President Trump, these tensions have escalated. The U.S. has publicly embraced the Political Declaration on Responsible Military Use of Artificial Intelligence and Autonomy, seeking to establish norms around ethical AI deployment in warfare.

At the same time, Trump's administration has taken an aggressive stance on AI as a strategic asset—particularly in its competition with China. By casting AI development as a national security priority, Trump promotes rapid innovation and military integration, invoking comparisons to the Manhattan Project.

Ethical Concerns and Rapid Militarization

Trump's Executive Order 14179, along with related defense directives, eliminates key regulatory barriers and expands the use of AI in military operations—including nuclear strategy. These actions have raised red flags among ethics and security experts.

The administration's loosened constraints on airstrikes and special operations (LaPorta, 2025) grant military commanders broad discretion, but at the cost of oversight and risk management. Without clear accountability frameworks, the potential for unintended consequences—from civilian casualties to AI misfires—grows substantially.

AI's integration into modern warfare is not inherently dangerous—but its deployment without adequate safeguards can destabilize democratic norms and global security. From battlefield robotics to virtual training simulations and autonomous weapons systems, innovation must be matched with ethical clarity and international cooperation.

The challenge now is not just how fast we can advance—but how wisely.

Cybersecurity Threats to National Security

State-sponsored cyberattacks represent a growing threat to national security, posing significant risks to critical infrastructure. These attacks, often carried out with the backing or tacit approval of hostile nation-states, aim to disrupt and infiltrate essential systems such as energy grids, financial networks, and water supplies. In response, nations must develop strategic defensive measures to safeguard their infrastructure and ensure public safety. This includes bolstering cybersecurity defenses through improved threat detection, intrusion prevention systems, and proactive incident response plans (*Secure Cyberspace and Critical Infrastructure*, n.d.).

These state-sponsored cyber threats are increasing in sophistication, which demands robust public-private partnerships. Government entities are called to collaborate with private industries and technology providers. Such collaboration is crucial in enhancing the resiliency of the diverse supply networks. Investing in cutting-edge cybersecurity technologies, such as artificial intelligence-driven threat detection and blockchain for secure data transactions, can significantly mitigate risks posed by cyber adversaries (McCaul, 2014).

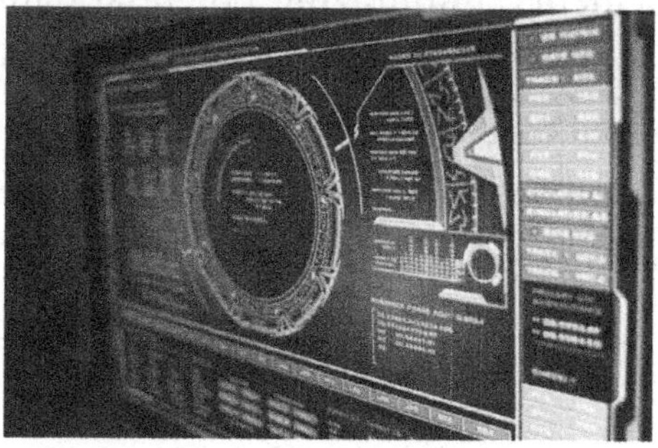

The Importance of Prevention and Previsibility

Besides resilience, governments work on consolidating their systems' protection. For this, they focus on cybersecurity awareness programs and a culture of security awareness within organizations. Insider threats remain a persistent challenge that organizations face from disgruntled employees or those susceptible to coercion. Some vital strategies are regular training sessions, ethical monitoring practices, and creating an environment where employees feel accountable for safeguarding sensitive information. Organizations should also implement protocols for flagging suspicious activities and encourage reporting procedures to preempt potential breaches.

To effectively safeguard infrastructure, countries must establish physical and digital security measures. These guidelines shall serve as roadmaps for best practices, providing clear directives on implementing security measures that enhance resilience against attacks and increase protection. For example, they can outline how to conduct regular risk assessments, prioritize vulnerabilities, and allocate resources efficiently to bolster defenses. Adhering to these guidelines ensures that stakeholders systematically address cybersecurity challenges, staying ahead of evolving threats.

Moreover, guidelines can streamline communication between government agencies and private sectors for prompt dissemination of threat intelligence. This collaborative approach enables swift responses to emerging cyber incidents, minimizing potential damages and disruptions. Standardized practices and guidelines simplify complex cybersecurity processes, making it easier for organizations to comply with regulations and protect their critical assets in unprecedented situations and the outburst of emergencies.

Trump's Approach to Cybersecurity

President Trump's second term will greatly affect U.S. cybersecurity policies, especially with the potential of executive orders that focus on the military's use of AI and offensive cyber operations. Understanding the ongoing threat of cyberattacks, particularly from countries like China and Iran, Trump's administration seems ready to take a more forceful approach. They plan to make it easier for military and intelligence agencies to carry out cyber operations that attack. This supports a "defend forward" strategy, which focuses on taking preemptive action against potential cyber threats and introduces a shift from defense to offense in cybersecurity.

However, this aggressive approach raises critical concerns regarding the implications for civil liberties and the potential for misuse of surveillance powers against political opponents and marginalized communities. The backdrop of a National Security Presidential Memorandum (NSPM) exemplifying this reveals a broader strategy to leverage national security for political objectives.

Additionally, the administration's inclination to repeal regulations that promote accountability raises alarms about the erosion of checks and balances within federal agencies, creating an environment ripe for overreach. As Trump seeks to position the US as a global leader in cybersecurity capabilities, the necessity of balancing these advancements with the safeguarding of individual rights and democratic values becomes increasingly urgent, setting the stage for a contentious and transformative era in U.S. cybersecurity policy.

Role of Technology in Border Security

The transformation of border security measures vividly demonstrates the impact of emerging technologies on military operations and national security. One of the most significant advancements is the deployment of surveillance technologies such as drones and cameras at borders. These tools have significantly enhanced the monitoring capacity, allowing authorities to observe hard-to-reach areas and respond swiftly to potential breaches. However, these enhancements come with a trade-off: the erosion of privacy rights. As noted by critics, the implementation of a "virtual wall" along borders can inadvertently create a "digital dumpster fire" for human rights and civil liberties (*Border Surveillance Technology*, n.d.).

Biometric systems are another streamlined identification process used at borders. Technologies like facial recognition and fingerprint scanning are becoming standard practices that facilitate the efficient processing of travelers. Yet, they also carry inherent risks, particularly bias against certain racial groups, which could further entrench discrimination and infringe upon civil liberties (Murdza, 2022).

Data collection strategies employed at borders enhance decision-making capabilities but demand meticulous management to ensure a balance between security objectives and individual rights. Large datasets gathered through various entry points allow for more informed policymaking and risk assessment.

Surveillance technologies operate in a complex space where necessity meets ethical considerations. Drones and cameras offer eyes in the sky but raise profound questions regarding constant supervision's psychological impact on communities living along borders. Maintaining a balance between effective security and preserving human dignity requires ongoing dialogue and policy reviews—emphasized by organizations advocating for migrant rights across Latin America, like the SinVigilancia coalition.

Furthermore, the use of robots, such as those tested by the U.S. Customs and Border Patrol, demonstrates how technology continues to redefine border patrol operations. While these innovations promise greater reach across rugged terrains, they simultaneously trigger alarms over potential civil liberty violations.

Trump's Policy on the National Borders

Under President Trump, the approach to utilizing AI and technology for border security is set to escalate significantly, emphasizing a militarized strategy that reflects his administration's broader goals. Central to this strategy is the implementation of the Political Declaration on Responsible Military Use of AI and the potential for aggressive offensive tactics aimed at immigration control. This aligns with Trump's longstanding rhetoric framing immigration as an "invasion," which has underpinned a push for heightened surveillance and technological measures to monitor and manage border crossings (Chishti & Bush-Joseph, 2025). The Biden administration's focus on responsible AI governance is anticipated to be dismantled, with Trump signaling intentions to repeal existing regulations that ensure safety and ethical considerations in technology use.

The proposed expansion of surveillance technologies will likely include AI-driven systems such as drone surveillance, biometric data collection, and advanced monitoring capabilities that can potentially blur the lines between security and civil liberties. Companies developing these technologies, like RealEye and Bavovna AI, are set to play pivotal roles in shaping the future of U.S. border security efforts.

RealEye is an emerging technology company that has developed AI-driven platforms designed to enhance border security through sophisticated monitoring and vetting processes. Its systems, named Masad and Fortress, provide real-time analysis of individuals entering the country, whether legally or otherwise, and allow for "semi-active" monitoring of their activities. This capability is especially significant in the context of U.S. border security, as it aims to identify and track individuals with a history of illegal or suspicious activities, thus bolstering national security measures.

Under Trump's administration, RealEye's technologies are expected to play a vital role in the heightened surveillance initiatives along the U.S.-Mexico border. The incorporation of such AI-driven solutions aligns with the administration's strategy to implement aggressive tactics for managing immigration and enhancing military capabilities at the border.

RealEye's deployment at U.S. borders will necessitate careful consideration of the implications for human rights amidst the overarching goal of strengthening national defense. This aggressive posture raises critical concerns about human rights and the treatment of migrants, as increased surveillance often leads to dangerous and exclusionary practices. As Trump's administration takes on this path, the focus on leveraging AI for border security will necessitate scrutiny regarding the implications for individual rights and the ethical use of technology in a democratic society.

Bringing It All Together

- AI and military tech are reshaping national defense—autonomous drones, cyber units, and sensor networks are making strategy faster, broader, and more complex.
- These technologies increase efficiency and reach, but also create new vulnerabilities—like cyberattacks, data leaks, and algorithmic bias in autonomous weapons.
- Trump's executive orders mark a shift toward a militarized approach to border security and immigration, emphasizing surveillance and AI enforcement tools.
- This reflects Trump's broader national security narrative, where policy decisions often prioritize control over civil liberties.
- The use of AI at the border raises serious concerns about privacy, discrimination, and abuse of power, especially when oversight is reduced.
- AI in border enforcement acts as a test case for how democracies balance innovation with ethics.
- Public conversations are now shifting toward privacy, accountability, and citizen rights, asking where to draw the line between protection and overreach.
- Achieving a true balance between security and liberty requires:
    - Transparent policymaking
    - Clear legal frameworks
    - Ongoing public engagement

- For readers, these debates aren't abstract—they influence daily life, personal privacy, and civic responsibility.
- The central question: *How should we respond when technology touches both national safety and individual freedom?*

Reader Reflection

The Leadership Dilemma

This final activity invites you to step into the shoes of decision-makers at the frontlines of ethical leadership. There are no "right" or "wrong" answers - only values, priorities, and trade-offs. Inspired by real-world decisions in AI and national defense, this exercise helps readers, leaders, and engaged citizens examine the tension between hardline security strategies and principled governance.

You'll explore two contrasting leadership paths:

- **Pathway 1 – Trump Response**: Prioritizes national security, dominance, and fast action.
- **Pathway 2 – Ethical Response**: Prioritizes transparency, human rights, and long-term trust.

Instructions:

Choose three scenarios from the following table and make a choice of action for each. Consider the following questions for each action you have chosen:

1. What leadership values influenced your choice?
2. What risks are you willing to accept to uphold these values?
3. How would you explain your decision to the public or your stakeholders?

Scenario	Trump Response (Security First)	Ethical Response (Principle First)
1. Predictive AI for Border Crossings	Deploy AI to profile travelers based on behavioral data, even if it risks bias.	Use AI for logistics only, avoiding personal profiling to prevent discrimination.
2. Automated Drone Surveillance	Equip drones with facial recognition for 24/7 border monitoring.	Limit drone use to environmental tracking, with oversight of human surveillance.
3. AI-Enhanced Military Decision Systems	Let autonomous systems recommend strike targets to reduce latency.	Require human-in-the-loop review for every critical decision.
4. AI in Disinformation Warfare	Use generative AI to counter foreign propaganda with synthetic media.	Stick to authentic diplomacy and verified facts in public messaging.
5. Facial Recognition at National Checkpoints	Mandate biometric scans nationwide for secure access points.	Offer opt-in ID programs with strict data governance and transparency.
6. Ethics Override Protocol	Allow emergency powers to bypass ethics during crises.	Preserve ethical guardrails even in emergencies, relying on diplomacy and law.
7. Recruitment Algorithms for Military Roles	Use AI to predict loyalty and filter high-risk candidates.	Train human recruiters to assess potential holistically, avoiding algorithmic bias.
8. AI Models Trained on Interrogation Data	Use real interrogation footage to train machine learning models.	Limit training to simulated, consensual data to avoid ethical violations.
9. Border Tech Contracting with Private Firms	Fast-track contracts with private AI vendors, even if unvetted.	Require independent review and transparency before awarding contracts.
10. Militarizing Civilian AI Projects	Repurpose civilian AI systems for defense without public notice.	Mandate public hearings and ethical audits before converting civilian tech.

Closing Thought

Leadership in AI isn't just about control, capability, or competition - it's about values. The real question is: When the stakes are high, will we choose power or principle?

The future will be shaped by those who dare to ask not just what we ***can*** do with technology, but what we ***should*** do.

Conclusion

At the heart of this book lies a singular purpose: to equip you with not just knowledge, but with agency - the confidence and awareness to participate meaningfully in America's (and global) ongoing dialogue about technology, power, and politics. The digital age is evolving at an extraordinary pace, redrawing boundaries and reshaping norms that once felt familiar. In these pages, we have aimed to move beyond headlines and sound bites, providing you with the context, analytical tools, and frameworks necessary to understand and responsibly navigate the changing landscape. By engaging with these themes, you've taken a significant step towards becoming not only an informed observer but also an active, ethical participant in society's most consequential conversations.

Throughout this journey, key patterns have emerged—sometimes subtle, sometimes stark—that reveal the intricate ways technology both empowers individuals and consolidates control. We have traced how regulatory choices, including pivotal executive orders and landmark legislation, can open doors for innovation or strengthen institutional safeguards, depending on their design and intent. The Trump administration's approach to deregulation underscored enduring questions about where the line should be drawn between enabling entrepreneurial dynamism and maintaining accountability in public life. Likewise, rapidly advancing technologies—from artificial intelligence to blockchain-based currencies—offer immense opportunities but also raise profound challenges, particularly as political actors leverage them both to engage voters and to shape policy.

One theme we returned to repeatedly is the dual character of technology as both tool and temptation. Digital platforms allow for unprecedented participation in democratic processes, yet they can also become instruments of manipulation. Political campaigns increasingly harness data analytics and AI to reach voters while blurring the lines between persuasion and intrusion. Cryptocurrency, too, destabilizes traditional financial influence channels even as it offers new forms of micro-donations and decentralized participation. Misinformation, amplified by algorithmic feeds and bot networks, can distort collective understanding and erode trust, threatening the shared reality that underpins effective self-governance.

These interwoven threads suggest a dynamic, and often uneasy, relationship between technological innovation and societal oversight. For every safeguard put in place, new vulnerabilities emerge; for each leap forward in efficiency or access, new ethical dilemmas arise. The ongoing debate over surveillance technologies in law enforcement and military contexts illustrates how difficult it can be to balance national security with civil liberties. The nuanced analysis of these case studies throughout the book aims to cultivate your awareness not only of current policy debates, but also of their underlying values and potential consequences.

Some of the most transformative insights from this exploration are neither simple nor comfortable. For example, the distinction between targeted messaging and psychological manipulation is rarely clear-cut. As digital strategies become smarter and more granular, traditional notions of informed consent and autonomy are being stretched in unexpected ways. Recognizing this ambiguity is vital—both for personal vigilance and for contributing thoughtfully to wider policy discussions. Another crucial revelation is the value of diverse voices in shaping technology's future: perspectives from marginalized communities, often overlooked in elite discourse, frequently point out blind spots and unintended outcomes, enriching the policymaking process.

Perhaps the most important 'aha' moment is realizing that innovation and oversight are not adversaries but partners in a healthy democracy. Embracing this tension allows us to resist false dichotomies—such as framing regulation solely as a brake on progress or innovation as an unqualified good. Many of the examples presented here highlight that durable solutions require compromise, humility, and above all, openness to multiple points of view. This mindset invites ongoing inquiry rather than fixed answers, encouraging readers to see themselves as lifelong learners and contributors to the public good.

If you've made it to this point, your commitment deserves acknowledgment. Navigating the intersections of technology, governance, and ethics is neither easy nor straightforward. You have engaged with complex material—wrestling with technical concepts, grappling with shifting political realities, and questioning assumptions about private and public life. This intellectual investment sets you apart from the vast majority who passively accept the status quo. Your willingness to examine not only what is happening, but also why and how it matters, is the foundation of any meaningful civic engagement.

This growth is not just academic; it is practical and deeply consequential. In an era when digital issues saturate our daily lives, your increased literacy allows you to better assess sources, challenge simplistic narratives, and participate in public debates with nuance and care. These are not minor skills—they are essential to safeguarding democracy, promoting equity, and fostering a culture of respectful disagreement. By choosing to think critically, you play a pivotal role in elevating the quality of dialogue both within your own circles and in the broader community.

The end of this book need not be the end of your exploration. Technology and its political entanglements will continue to evolve—often in unpredictable directions. If you found particular topics compelling, consider delving deeper into included resources, recommended organizations, and ongoing case studies highlighted throughout the text. Attending public forums, reading recent policy briefs, or joining online seminars are accessible ways to extend your learning. Even informal activities, such as discussing these issues with family or colleagues, reinforce your understanding and expose you to alternative viewpoints.

Active citizenship in the digital era is an ongoing practice rather than a set destination. Consider ways you might contribute to shaping the conversations and policies that affect your world, whether by sharing insights on social media, volunteering for tech-focused advocacy groups, or participating in local government initiatives. If you feel strongly about a particular concern discussed here—be it AI transparency, data privacy, or equitable internet access—small actions can aggregate into meaningful change. Sometimes, simply asking the right question in a crowded room can shift debate. How might your voice enrich and diversify the emerging consensus around these critical issues?

Building connections with others who share your curiosity multiplies your impact. Communities—both virtual and in-person—play a crucial role in driving sustained attention to the ethical dimensions of technology. Joining or initiating reading groups, public panels, or online forums creates spaces where knowledge is not only accumulated but also challenged and refined. Through these networks, your individual efforts align with those of others who are equally committed to steering digital innovation toward the common good. What role could you assume in helping your community grapple with the trade-offs inherent in technological progress?

Above all, thank you for investing your time, energy, and trust in this book's project. Your engagement demonstrates that careful, sober consideration of technology's possibilities and risks remains both necessary and possible amidst the noise of contemporary discourse. You are now part of a larger, ongoing effort to ensure that American and global democracy and innovation evolve together sustainably and justly. The path forward will demand vigilance, curiosity, and empathy, but it will also be shaped by the cumulative influence of readers like you, willing to think critically and act collaboratively. Together, we can help guide technology toward serving, rather than dominating, the society we all share.

A Small Favour That Makes A Big Difference

If this book has spoken to you, I'd be deeply grateful if you could take a moment to leave a review on Amazon.

Your reflections do more than add stars — they help other readers decide if this book is right for them, and they keep these ideas alive by sparking new conversations. Reviews also help authors like me continue to grow and share more resources with the people who need them most.

Thank you for lending your voice. Your support means more than you know.

Kai Denham

Kai Denham is a writer, researcher, and observer of technology's evolving role in society. With a curiosity for what lies just beyond the edge of innovation, his work bridges the technical and the human. Kai writes about where technology meets real life—decoding AI, crypto, and the shifting digital world so the rest of us don't have to.

With a knack for turning research into something readable, his work highlights what matters, why it matters, and what might come next. Equal parts curious and clear-headed, Kai's voice invites readers to think critically, stay informed, and maybe even take action.

References

$LIBRA scandal: Argentina's judiciary opens investigation into Milei. (2025, February 21). Buenos Aires Herald. https://buenosairesherald.com/politics/judiciary/libra-scandal-argentinas-judiciary-opens-investigation-into-milei

2024 global forum on building trust and reinforcing democracy. (2024). OECD. https://www.oecd.org/content/dam/oecd/en/about/programmes/reinforcing-democracy-initiative/2024-OECD-Global-Forum-Key-Issues-Paper.pdf

Actions for transparent and accountable digital governance. (n.d.). Open Government Partnership. https://www.opengovpartnership.org/actions-for-transparent-and-accountable-digital-governance

Adams, D.G., Gerratana, F.L., & Effron, S. (2025, February 20). A new era for crypto regulation & innovation? The crypto executive order, a rebooted SEC crypto task force & the journey ahead. *The National Law Review, 15*(76). https://natlawreview.com/article/new-era-crypto-regulation-innovation-crypto-executive-order-rebooted-sec-crypto

Adrian, T., He, D., Ismail, A., & Moretti, M. (2023, July 18). Crypto needs comprehensive policies to protect economies and investors. *IMF Blog.* https://www.imf.org/en/Blogs/Articles/2023/07/18/crypto-needs-comprehensive-policies-to-protect-economies-and-investors

AI at war: The next revolution for military and defense. (2023, August 4). *Capitology Blog: Capitol Technology University.* https://www.captechu.edu/blog/ai-war-next-revolution-military-and-defense

Akselrod, O. & Venzke, C. (2025, February 11). *Trump's efforts to dismantle AI protections, explained.* ACLU. https://www.aclu.org/news/privacy-technology/trumps-efforts-to-dismantle-ai-protections-explained

Alessa, M. (2024, July). *The impact of fintech on the financial services industry.* ResearchGate. https://www.researchgate.net/publication/382456597_The_Impact_of_Fintech_on_the_Financial_Services_Industry

Alotaibi, E. M., Issa, H., & Codesso, M. (2025). Blockchain-based conceptual model for enhanced transparency in government records: A design science research approach. *International Journal of Information Management Data Insights, 5*(1). https://doi.org/10.1016/j.jjimei.2024.100304

Alto, D. (2024, June 1). Cryptocurrency regulation: Balancing innovation and investor protection. *Medium.* https://medium.com/@davidalto/cryptocurrency-regulation-balancing-innovation-and-investor-protection-e4388e43474f

American Library Association. (2018). Evaluating Sources: Criteria and Resources.

American Library Association. (2018). Guide to Fake News and Misinformation.

American Press Institute. (2019). Rebuilding Public Trust in News.

Aratani, L. (2025, February 3). *Trump signs order to create sovereign wealth fund that could buy TikTok.* The

Guardian. https://www.theguardian.com/us-news/2025/feb/03/trump-sovereign-wealth-fund-tiktok

Asimakopoulos, G., Antonopoulou, H., Giotopoulos, K., & Halkiopoulos, C. (2025, February 18). Impact of information and communication technologies on democratic processes and citizen participation. *Societies, 15*(2), 40. https://doi.org/10.3390/soc15020040

Associated Press. (2024, March 26). *What to know about Truth Social, Trump's social media platform.* PBS News. https://www.pbs.org/newshour/politics/what-to-know-about-truth-social-trumps-social-media-platform

Australian Government. (2010). *Freedom of Information Amendment (Reform) Act 2010.* Federal Register of Legislation. https://www.legislation.gov.au/C2010A00051/latest

Bank for International Settlements. (2025). Stablecoins and the quest for global liquidity (BIS Quarterly Review, June). https://www.bis.org/publ/qtrpdf/r_qt2506e.pdf

Bloomberg. (2025, June 12). SEC drops key crypto rules in major regulatory rollback under Trump administration. Bloomberg.com. https://www.bloomberg.com/news/articles/2025-06-12/sec-drops-key-crypto-rules-decentralized-finance

Border surveillance technology. (n.d.). Electronic Frontier Foundation. https://www.eff.org/issues/border-surveillance-technology

Bose, N. & Ward, J. (2025, March 7). *Trump signs order to establish strategic bitcoin reserve.* Reuters. https://www.reuters.com/technology/trump-signs-order-establish-strategic-bitcoin-reserve-white-house-crypto-czar-2025-03-07

Bose N., McGee, S. & Conlin, M. (2025, March 7). *Crypto leaders meet at Trump's summit with strategic reserve in focus.* Reuters. https://www.reuters.com/technology/crypto-leaders-meet-trumps-summit-strategic-reserve-focus-2025-03-07

Bradbury, D. (2025, January 16). *What will cybersecurity look like under Trump 2.0?* ISMS. https://www.isms.online/cyber-security/what-will-cybersecurity-look-like-under-trump-2-0

Carrillo, P.E., Mackowski, M.J., Maschek, W.A., Goldstein, B., Jacobson, J.B., Friel, A.L., & Kirk, M. (2025, February). *Key insights on President Trump's new AI executive order and policy & regulatory implications.* Squire Patton Boggs. https://www.squirepattonboggs.com/en/insights/publications/2025/02/key-insights-on-president-trumps-new-ai-executive-order-and-policy-regulatory-implications

Chen, M. A., Wu, Q., & Yang, B. (2019, April 4). How valuable is FinTech innovation? *The Review of Financial Studies, 32*(5), 2062-2106. https://doi.org/10.1093/rfs/hhy130

Chheda, H. (2024, December 10). Trump's approach to cybersecurity policies. *Sprinto.* https://sprinto.com/blog/trumps-approach-to-cybersecurity-policies

Chishti, M. & Bush-Joseph, K. (2025, January 23). *With "shock and awe," the second Trump term opens with a bid to strongly reshape immigration.* MPI. https://www.migrationpolicy.org/article/trump-second-term-begins-immigration

Clarke, C. (n.d.). *Deconstructing the deep state.* Government Executive. https://www.govexec.com/feature/gov-exec-deconstructing-deep-state/

Clarke, L. (2025, January 2). *'Nobody was tricked into voting for Trump': Why the disinformation panic is over.* Politico. https://www.politico.eu/article/nobody-tricked-vote-donald-trump-disinformation-panic-over

CoinDesk. (2025, July 17). Trump signs stablecoin bill into law. CoinDesk. https://www.coindesk.com/policy/2025/07/17/trump-signs-stablecoin-bill-into-law/

Coinmetro Editorial Team. (2024, August 21). *Regulatory sandboxes: Fostering crypto innovation within legal frameworks.* Coinmetro. https://coinmetro.com/learning-lab/regulatory-sandboxes

Cole, D. (2025, January 22). *Trump's executive order for TikTok shows his willingness to skirt federal laws and keep app in limbo, legal experts say*. CNN. https://edition.cnn.com/2025/01/22/politics/tiktok-trump-legal-executive-order/index.html

Cook, J., & Lewandowsky, S. (2020). The Debunking Handbook. Yale University.

Council of Europe. (2019). National Strategies for Countering Disinformation.

Council on Foreign Relations. (2020). A Guide to Spotting Fake News.

Cousens, M. (2024, December 2). *2024 Post-election survey: A majority of new Trump voters used social media as main news source.* Navigator. https://navigatorresearch.org/2024-post-election-survey-a-majority-of-new-trump-voters-used-social-media-as-main-news-source

Crypto and digital assets: Regulatory challenges. (n.d.). KPMG. https://kpmg.com/us/en/articles/2022/ten-key-regulatory-challenges-2022-crypto-digital-assets.html

Csernatoni, R. (2024, December 18). *Can democracy survive the disruptive power of AI?* Carnegie Endowment for International Peace. https://carnegieendowment.org/research/2024/12/can-democracy-survive-the-disruptive-power-of-ai

Deakin University. (2024, January 29). *Australia's FOI Act not fit for purpose in age of automation*. https://www.deakin.edu.au/about-deakin/news-and-media-releases/articles/australias-foi-act-not-fit-for-purpose-in-age-of-automation

Del Sarto, N., Ramirez, I. C., & Gai, L. (2025). *Impacts of FinTech funding announcements on traditional banks: An event study analysis. Journal of Economics and Business, 133*, 106231. https://doi.org/10.1016/j.jeconbus.2024.106231

De Mott, F. (2025, February 28). *'Crypto president' no more? Trump's tariffs have helped wipe out $1 trillion in value since December.* Business Insider. https://www.businessinsider.com/bitcoin-price-crypto-market-cap-donald-trump-tariff-inflation-outlook-risk-2025-2

Duffy, C. (2025, January 21). *Trump announces a $500 billion AI infrastructure investment in the US*. CNN. https://edition.cnn.com/2025/01/21/tech/openai-oracle-softbank-trump-ai-investment/index.html

Dunn, G. (2025a, February 3). *Impact of Trump administration regulatory freeze memorandum: Selected regulations and agency actions*. Gibson Dunn. https://www.gibsondunn.com/impact-of-trump-administration-regulatory-freeze-memorandum-selected-regulations-and-agency-actions

Dunn, G. (2025b, February 4). *Two weeks in: Key Trump administration developments in tech policy*. Gibson Dunn. https://www.

gibsondunn.com/two-weeks-in-key-trump-administration-developments-in-tech-policy

Ermey, R. (2025, January 31). *3 ways Trump has impacted the crypto market so far: Everything on the wish list 'pretty much happened'.* CNBC. https://www.cnbc.com/2025/01/31/how-trump-has-impacted-crypto-so-far.html

European Commission. (2018). Tackling Online Disinformation: A European Approach.

European Digital Media Observatory. (2020). Handbook for Journalists.

Faure, V. (2024, November 15). *How America's tech right came to power*. Le Monde. https://www.lemonde.fr/en/international/article/2024/11/15/how-america-s-tech-right-came-to-power_6732989_4.html

Fechner, H., Johnson, J., McMurrough, M., Huffman, R., Shapanka, M. & Gweon, A. (2025, February 7). *Trump administration seeks public comment on AI action plan*. Covington. https://www.globalpolicywatch.com/2025/02/trump-administration-seeks-public-comment-on-ai-action-plan

Financial Times. (2025, July 18). Trump set to open $9 trillion 401(k) pensions market to crypto: FT. FT.com. https://www.ft.com/content/crypto-401k-ft-report

Friedman, B. & Delerme, R. (2025, February 21). *We study mass surveillance for social control, and we see Trump laying the groundwork to 'contain' people of color and immigrants.* The Conversation. https://theconversation.com/we-study-mass-surveillance-for-social-control-and-we-see-trump-laying-the-groundwork-to-contain-people-of-color-and-immigrants-221073

Gerken, T., McMahon & Rahman-Jones, I. (2025, January 22). *What does Trump's executive order mean for TikTok and who might buy it?* BBC. https://www.bbc.com/news/articles/clyng762q4eo

Giles, C. (2025, February 5). *Trump's data deletions pose a stark threat to public health.* The Bureau of Investigative Journalism. https://www.thebureauinvestigates.com/stories/2025-02-05/trumps-data-deletions-pose-a-stark-threat-to-public-health

Gill, J. (2024, October 9). AI ethics: The ethical landscape of artificial intelligence. *Xenonstack.* https://www.xenonstack.com/blog/ethics-artificial-intelligence

Governing with artificial intelligence: Are governments ready? (2024, June). *OECD Artificial Intelligence Papers, (20).* https://www.oecd.org/content/dam/oecd/en/publications/reports/2024/06/governing-with-artificial-intelligence_f0e316f5/26324bc2-en.pdf

Government surveillance. (n.d.). United States Congresswoman Zoe Lofgren. http://lofgren.house.gov/issues/innovation-and-technology/government-surveillance

Gracia, S. (2024, September 19). *Americans in both parties are concerned over the impact of AI on the 2024 presidential campaign.* Pew Research Center. https://www.pewresearch.org/short-reads/2024/09/19/concern-over-the-impact-of-ai-on-2024-presidential-campaign

Gross, P. (2025, February 24). *How Trump's tech agenda could impact hiring this year.* Missouri Independent. https://missouriindependent.com/2025/02/24/repub/how-trumps-tech-agenda-could-impact-hiring-this-year

Guiding principles on government use of surveillance technologies. (2023, March). Freedom Online Coalition. https://freedomonlinecoalition.com/guiding-principles-on-government-use-of-surveillance-technologies

Haddad, C. (2025, February 12). *Fintech and the criticality of compliance*. Citi. https://www.citigroup.com/global/insights/fintech-and-the-criticality-of-compliance

Harris, S. (2025, January 23). *Trump's first shot in his war on the 'Deep State.'* The Atlantic. https://www.theatlantic.com/politics/archive/2025/01/trump-executive-order-security/681423

Hayes, A. (2025, January 10). *Net neutrality overturned: Appeals court ruling redraws the Internet's future*. Investopedia https://www.investopedia.com/the-evisceration-of-net-neutrality-8770956

Hirschfeld, A. (2025, February 10). *Trump accused of leading a 'multipronged' attack on US media*. Aljazeera. https://www.aljazeera.com/economy/2025/2/10/trump-accused-of-leading-a-multipronged-attack-on-us-media

Hotaling, G. (2024, November). *As Gensler departs, here's what a second Trump term means for investment firms*. Confluence. https://www.confluence.com/new-in-regtech-november-2024-what-a-second-term-for-trump-means-for-investment-firms

House Committee on Financial Services. (2025, July 10). Section-by-section: Guiding and Establishing National Innovation for U.S. Stablecoins Act (GENIUS Act). U.S. House of Representatives. https://financialservices.house.gov/uploadedfiles/genius_act_section_by_section.pdf

Howe, E. G., & Elenberg, F. (2020, October 1). Ethical challenges posed by big data. *Innovations in Clinical Neuroscience*,

17(10-12):24–30. https://pmc.ncbi.nlm.nih.gov/articles/PMC7819582

How US regulatory consolidation could reshape the RegTech landscape. (2025, February 18). Fintech Global. https://fintech.global/2025/02/18/how-us-regulatory-consolidation-could-reshape-the-regtech-landscape

Huelss, H. (2025, February 18). *Google's brave new world? Big tech, military AI, and the Trump effect*. AutoNorms. https://www.autonorms.eu/googles-brave-new-world-big-tech-military-ai-and-the-trump-effect

InCountry Staff. (2025, February 11). *Trump's impact on global data sovereignty.* InCountry. https://incountry.com/blog/trumps-impact-on-global-data-sovereignty

International Fact-Checking Network. (n.d.). Code of Principles. Poynter Institute.

Jacobson, J., Swanson, S., Helleputte, C., & Emery, R. (2025, February 11). *A new era: Trump 2.0 highlights for privacy and AI.* Squire Patton Boggs. https://www.privacyworld.blog/2025/02/a-new-era-trump-2-0-highlights-for-privacy-and-ai

Javaid, M., Haleem, A., Singh, R. P., Suman, R., & Khan, S. (2022, July). A review of Blockchain Technology applications for financial services. *BenchCouncil Transactions on Benchmarks, Standards and Evaluations, 2*(3). https://doi.org/10.1016/j.tbench.2022.100073

Karkara, R. (2025, February 8). *US crypto policies under Trump: A new era for global innovation*. Economic Times. https://economictimes.indiatimes.com/markets/cryptocurrency/

us-crypto-policies-under-trump-a-new-era-for-global-innovation/articleshow/118055732.cms?from=mdr

Katz, J. & Urby, E. (2025, February 13). *Dangerous cracks in US democracy pillars*. Brookings. https://www.brookings.edu/articles/dangerous-cracks-in-us-democracy-pillars/

Käufer, T. (2025, February 21). *Argentina's Milei faces credibility crisis over crypto scam*. DW. https://www.dw.com/en/argentinas-milei-faces-credibility-crisis-over-crypto-scam/a-71691738

Kavanagh, C. (2019, August 28). *New tech, new threats, and new governance challenges: An opportunity to craft smarter responses?* Carnegie Endowment for International Peace. https://carnegieendowment.org/2019/08/28/new-tech-new-threats-and-new-governance-challenges-opportunity-to-craft-smarter-responses-pub-79736

Kehinde, S., Simon-Ilogho, B., Kehinde, K. & Kehinde, T. (2024). *Exploring the impact of AI on voter confidence and election information in 2024*. Qeios. https://doi.org/10.32388/ut898q

Kerr, D. & Bhuiyan, J. (2025, February 22). *Crypto and big tech's backing pays off as Trump makes tech-friendly moves*. The Guardian. https://www.theguardian.com/technology/2025/feb/22/crypto-big-tech-trump

Khalili, J. (2025, February 28). *The SEC is abandoning its biggest crypto lawsuits*. Wired. https://www.wired.com/story/sec-is-giving-up-biggest-crypto-lawsuits

Kimpel, S.H. (2025, January 27). *President Trump issues executive order on digital assets*. Hunton. https://www.hunton.com/blockchain-legal-resource/president-trump-issues-executive-order-on-digital-assets

Kinnard, M. (2025, February 21). *A comprehensive look at DOGE's firings and layoffs so far.* AP. https://apnews.com/article/doge-firings-layoffs-federal-government-workers-musk-d33cdd7872d64d2bdd8fe70c28652654

Königs, P. (2022, February 5). Government surveillance, privacy, and legitimacy. *Philosophy & Technology, 35*(8). https://doi.org/10.1007/s13347-022-00503-9

Kortam, R. (2022, November 23). *Global co-operation key to developing effective cryptocurrency regulation*. OMFIF. https://www.omfif.org/2022/11/global-co-operation-key-to-developing-effective-cryptocurrency-regulation

LaChapelle, C., & Tucker, C. (2023, November 28). *Generative AI in political advertising.* Brennan Center for Justice. https://www.brennancenter.org/our-work/research-reports/generative-ai-political-advertising

Laghari, R. (2025, January 23). *Governance in the digital era: The role of technology in enhancing transparency and accountability*. Medium. https://medium.com/@riazleghari/governance-in-the-digital-era-the-role-of-technology-in-enhancing-transparency-and-accountability-78bd2ffe8208

Lang, H. & Hunnicutt, T. (2025, January 24). *Trump orders crypto working group to draft new regulations, explore national stockpile*. Reuters. https://www.reuters.com/business/finance/trump-signs-order-create-cryptocurrency-working-group-2025-01-23

LaPorta, J. (2025, February 28). *Politics Trump eases rules on military raids and airstrikes, expanding range of who can be targeted.* CBS. https://www.cbsnews.com/news/trump-eases-rules-military-raids-airstrikes-targets

Lee, M. (2025, February 5). *Protesters in cities across the US rally against Trump's policies, Project 2025 and Elon Musk.* AP. https://apnews.com/article/50501-protests-project-2025-trump-state-capitols-ddd341171a54ba9b498cbfe7530e18ab

Lewandosky, S. (2025, February 6). Free speech, fact checking, and the right to accurate information. *Science, 387*(6734). https://www.science.org/doi/10.1126/science.adv4632

Library of Congress. (n.d.). *Cryptocurrency & blockchain technology*. https://guides.loc.gov/fintech/21st-century/cryptocurrency-blockchain

Lu, W. (2024, October 3). Inevitable challenges of autonomy: ethical concerns in personalized algorithmic decision-making. *Humanities and Social Sciences Communications, 11*(1321). https://doi.org/10.1057/s41599-024-03864-y

Majumdar, S., Moussawi, R. & Yaylacicegi, U. (2006, August 15). *The impact of mergers on performance in the US telecommunications industry, 1988 to 2001*. ResearchGate. https://www.researchgate.net/publication/256029091_The_Impact_of_Mergers_on_Performance_in_the_US_Telecommunications_Industry_1988_to_2001

Mandell, J. & Tesfaye, R. (2025, February 24). *AI trends for 2025 - AI and financial services: Balancing innovation with regulatory compliance.* JD Supra. https://www.jdsupra.com/legalnews/ai-trends-for-2025-ai-and-financial-1976406

Mariah Wildgen, "New Trump Administration AI Guidance Leaves Civil Rights Behind," *Leadership Conference on Civil and Human Rights*, April 14 2025, https://civilrights.org/blog/new-trump-administration-ai-guidance-leaves-civil-rights-behind/ .

Matthias, M. (n.d.). Truth Social. In *Encyclopedia Britannica*. Retrieved on March 15, 2025. https://www.britannica.com/topic/Truth-Social

McCaul, M. T. (2014, July 29). *H.R.3696 - 113th Congress (2013-2014): National cybersecurity and critical infrastructure protection act of 2014*. Congress.gov. https://www.congress.gov/bill/113th-congress/house-bill/3696

McGreal, C. (2025, February 1). *Wrecking ball: Trump's power-hungry orders wage war on US government*. The Guardian. https://www.theguardian.com/us-news/2025/feb/01/trump-executive-order-power-grab

McKinsey & Company. (2024, January 16). *What is fintech?* https://www.mckinsey.com/featured-insights/mckinsey-explainers/what-is-fintech

McKinsey & Company. (2024). Global Banking Annual Review 2024: The $117 trillion opportunity in stablecoins. McKinsey.com. https://www.mckinsey.com/industries/financial-services/our-insights/global-banking-annual-review-2024

McMahon, L. Kleinman, Z. & Subramanian, C. (2025, January 7). Facebook and Instagram get rid of fact checkers. *BBC*. https://www.bbc.com/news/articles/cly74mpy8klo

MIT Media Lab, Center for Civic Media. (2019). Community Resilience Research.

Molnar, P. (2025a, January 7). *The walls have eyes: How Trump plans to supercharge border security with tech*. Bulletin of the Atomic Scientists. https://thebulletin.org/2025/01/the-walls-have-eyes-how-trump-plans-to-supercharge-border-security-with-tech

Molnar, P. (2025b, January 2). *Border technologies under Trump 2.0.* Just Security. https://www.justsecurity.org/105662/border-technologies-under-second-trump-administration

Monash University. (2024, June 21). *The need to improve freedom of information practices in Australia*. Monash Lens. https://lens.monash.edu/@politics-society/2024/06/21/1386820/the-culture-of-implementing-freedom-of-information-in-australia

Mortensen, O. (2025, January 28). *How many users does Truth Social have?* SEO.AI. https://seo.ai/blog/how-many-users-does-truth-social-have

Murdza, K. (2022, March 4). *New border surveillance technology raises privacy concerns and could increase deaths*. Immigration Impact. https://immigrationimpact.com/2022/03/04/border-surveillance-technology-privacy-deaths

Nelson, E. (2024, October 23). *Fintech compliance and how to maintain it*. Hyperproof. https://hyperproof.io/resource/fintech-compliance-and-how-to-maintain-it

News Literacy Project. (n.d.). Educator Resources. newslitproject.org..

Nicas, J. & Yaffe-Bellany, D. (2025, February 28). *Milei, $Melania and memecoins: Unraveling Argentina's crypto fiasco*. The New York Times. https://www.nytimes.com/2025/02/28/world/americas/argentina-crypto-scandal-president.html

Nickerson, R. S. (1998). Confirmation Bias: A Ubiquitous Phenomenon in Many Guises. Review of General Psychology, 2(2), 175–220.

Nieuchowicz, A. & Archer, J. (2025, February 5). *Trump Administration's executive order on digital assets: A significant shift in U.S. crypto policy.* Carlton Fields. https://www.carltonfields.com/insights/publications/2025/

trump-administrations-executive-order-on-digital-assets-a-significant-shift-in-u-s-crypto-policy

Shafiabady, N. & O'Neil A. (2025) America first, ethics second: The implications of Trump's AI Executive Order. *Lowy Institute.* https://www.lowyinstitute.org/the-interpreter/america-first-ethics-second-implications-trump-s-ai-executive-order .

O'Neill, B. (2025, February 28). *The president's declassification power is a double-edged sword.* Just Security. https://www.justsecurity.org/108514/presidents-declassification-double-edged-sword

Ochs, D., Betts, J.G., & Zagger, Z.V. (2025, February 7). *Trump Administration unveils new AI policy, reverses Biden's regulatory framework.* Ogletree Deakins. https://ogletree.com/insights-resources/blog-posts/trump-administration-unveils-new-ai-policy-reverses-bidens-regulatory-framework

Office of the Australian Information Commissioner. (n.d.). *Freedom of information.* https://www.oaic.gov.au/freedom-of-information

Office of the Victorian Information Commissioner. (2021). *Information Commissioner welcomes Parliamentary review of the FOI Act to modernise access to information.* https://ovic.vic.gov.au/mediarelease/information-commissioner-welcomes-parliamentary-review-of-the-foi-act-to-modernise-access-to-information

Olaniran, B., & Williams, I. (2020, February 27). *Social media effects: Hijacking democracy and civility in civic engagement.* Platforms, Protests, and the Challenge of Networked Democracy. https://doi.org/10.1007/978-3-030-36525-7_5

Oliphant, J. & Holland, S. (2025, February 18). *How Trump plans to cement control of government by dismantling the 'deep state.'* Reuters. https://www.reuters.com/world/us/

how-trump-plans-cement-control-government-by-dismantling-deep-state-2025-01-18

Patel, M. (2024, February 29). *The ethics of AI addressing bias, privacy, and accountability in machine learning*. CloudThat. https://www.cloudthat.com/resources/blog/the-ethics-of-ai-addressing-bias-privacy-and-accountability-in-machine-learning

Pearl, M. (2025, January 27). *Trump's moves to modernize U.S. technology policy*. CSIS. https://www.csis.org/analysis/trumps-moves-modernize-us-technology-policy

Lukens, P. (2025) How President Trump's AI Executive Order Could Reshape Policing. *Police1*. https://www.police1.com/artificial-intelligence/how-president-trumps-ai-executive-order-could-reshape-policing/ .

Political declaration on responsible military use of artificial intelligence and autonomy. (n.d.). U.S. Department of State. https://www.state.gov/bureau-of-arms-control-deterrence-and-stability/political-declaration-on-responsible-military-use-of-artificial-intelligence-and-autonomy

PolitiFact. (n.d.). Poynter Institute. politifact.com..

President Trump issues executive order to establish digital assets regulatory framework. (2025, February 10). O'Melveny. https://www.omm.com/insights/alerts-publications/president-trump-issues-executive-order-to-establish-digital-assets-regulatory-framework

Raffio, N. (2024, March 6). *Can we still trust the polls?* USC. https://today.usc.edu/can-we-still-trust-the-polls

Rasul, M. E., Halversen, A., & Smith, J. (2025). "When you're a star, they let you do it:" Trump, Twitter, and moral disengagement. *Communication and the Public*. https://doi.org/10.1177/20570473251314521

Rattanasevee, P., Akarapattananukul, Y. & Chirawut, Y. (2024, December 19). Direct democracy in the digital age: opportunities, challenges, and new approaches. *Humanities and Social Sciences Communications; Palgrave Macmillan, 11*(1681). https://doi.org/10.1057/s41599-024-04245-1

Reuters. (2025, March 7). Stablecoins will preserve the U.S. dollar's dominance, Bessent says. Reuters.com. https://www.reuters.com/technology/stablecoins-will-preserve-us-dollar-dominance-bessent-says-2025-03-07/

Safarli, G. J. & Safarli, A.J.. (2024). *The Impact of cryptocurrency adoption on traditional baking system: A theoretical study.* https://doi.org/10.61413/zisz5040

Saha, S., Bishwas, P. C., Das, U., & Arshi, A. (2024, April 26). Is fintech just an innovation? Impact, current practices, and policy implications of fintech disruptions. *International Journal of Economics, Business and Management Research, 8*(4). https://doi.org/10.51505/ijebmr.2024.8412

Saltini, A. (2025, January 30). *AI and nuclear strategy under Trump 2.0: What to expect.* Open Nuclear Network. https://platform.opennuclear.org/thoughtroom/external-contributions/ai-and-nuclear-strategy-under-trump-20-what-to-expect

Samples, J. (2019, April 9). *Why the government should not regulate content moderation of social media.* CATO Institute. https://www.cato.org/policy-analysis/why-government-should-not-regulate-content-moderation-social-media#the-presumption-against-public-regulation-of-social-media

Sandler, B.D., Markel, S.C., McKnight, A.M., James, R.A., & Ajanaku, D.F. (2025, January 30). *Trump 2.0: AI and data centers in a time of legal and technological disruption*. Pillsbury. https://www.pillsburylaw.com/en/news-and-insights/ai-data-centers-trump.html

Scaduto, F., Pozza, D., Harasic-Yaksic, V., Muhlendorf, K., Alli, K., Lerman, L. & Want, I. (2025, February 5). *President Trump's crypto czar outlines federal government's new approach to digital assets*. Wiley. https://www.wiley.law/alert-President-Trumps-Crypto-Czar-Outlines-Federal-Governments-New-Approach-to-Digital-Assets

Schmelzer, R. & Walch, K. (2025, January 15). *Top 10 ethical considerations for AI projects*. Project Manager Institute. https://www.pmi.org/blog/top-10-ethical-considerations-for-ai-projects

Schubert, K. D., & Barrett, D. (2024, May 25). Data governance, privacy, and ethics. In: Lacity, M.C., Coon, L. (eds) *Human Privacy in Virtual and Physical Worlds. Technology, Work and Globalization*. Palgrave Macmillan. https://doi.org/10.1007/978-3-031-51063-2_5

Schwab, K. (2016, January 14). The fourth industrial revolution: What it means, how to respond. World Economic Forum. https://www.weforum.org/stories/2016/01/the-fourth-industrial-revolution-what-it-means-and-how-to-respond

Schwarcz, S.L. (2024, December 6). Regulating financial innovation: FinTech, crypto-assets, DeFi, and beyond. *The Business Lawyer, 79*(3). https://www.americanbar.org/groups/business_law/resources/business-lawyer/2024-summer/regulating-financial-innovation-fintech-crypto-assets-defi-and-beyond

Scicchitano, M. (n.d.). *How cryptocurrencies may impact the banking industry*. Wolf & Company. https://www.wolfandco.com/resources/insights/how-cryptocurrencies-may-impact-the-banking-industry

Science Media Centre. (n.d.). How to Evaluate Scientific Claims.

Scoble, R. & Cronin, I. (2024, November 12). *AI in military applications*. Unaligned Newsletter. https://www.unaligned.io/p/ai-in-military-applications

Secure cyberspace and critical infrastructure. (n.d.) Department of Homeland Security. https://www.dhs.gov/secure-cyberspace-and-critical-infrastructure

Shahbaz, A. (2018). *The rise of digital authoritarianism*. Freedom House. https://freedomhouse.org/report/freedom-net/2018/rise-digital-authoritarianism

Shivaram, D. (2025, March 7). *Trump pledges support for crypto sector at White House summit*. NPR. https://www.npr.org/2025/03/07/nx-s1-5319049/trump-crypto-summit

Sixth circuit blocks net neutrality. (2025, January 6). Brownstein. https://www.bhfs.com/insights/alerts-articles/2025/sixth-circuit-blocks-net-neutrality

Spitznagel, E. (2025, January 18). *US news from robots to AI: The tech behind Trump's plan to secure the Southern Border*. New York Post. https://nypost.com/2025/01/18/us-news/robots-to-ai-the-technology-behind-trumps-plan-to-seal-the-southern-border

Squire Patton Boggs, "Key Insights on President Trump's New AI Executive Order and Policy & Regulatory Implications," February 2025, https://www.squirepattonboggs.com/-/media/files/insights/publications/2025/02/

key-insights-on-president-trumps-new-ai-executive-order-and-policy--regulatory-implications/president_trumps_new_ai_eo.pdf .

Stone, P. (2025, February 7). *Trump's blitz to expand his power is direct threat to democracy, experts say*. The Guardian. https://www.theguardian.com/us-news/2025/feb/07/trump-executive-actions-democracy

Subin, S. (2025, January 27). *Nvidia sheds almost $600 billion in market cap, biggest one-day loss in U.S. history*. CNBC. https://www.cnbc.com/2025/01/27/nvidia-sheds-almost-600-billion-in-market-cap-biggest-drop-ever.html

Suderman, A. (2025, February 17). *A look at the Trump crypto ventures that will help the industry–and likely enrich his family.* PBS News. https://www.pbs.org/newshour/politics/a-look-at-the-trump-crypto-ventures-that-will-help-the-industry-and-likely-enrich-his-family

Suderman, A. & Debre, I. (2025, February 21). *Argentina's crypto scandal dings Milei, involves strange mix of characters.* AP. https://apnews.com/article/argentina-milei-meme-coins-crypto-melania-e83b5ffd61b1dbc9e7c1272096d39aaa

Sudhini, R. (2024). AI Technology to Increase US Government Transparency. *STEPS: Science, Technology, Engineering, and Policy Studies, 9.* https://www.potomacinstitute.org/steps/images/PDF/Articles/24/SudhiniSTEPS9a.pdf

Swanholt, E. K., McGaver Sikora, K, Chester, M., & Mitcham, M.R.(2025). An executive branch for the people, by the people: What the new administration's executive orders mean for an independent CPSC. *The National Law Review, 15*(76). https://natlawreview.com/article/

executive-branch-people-people-what-new-administrations-executive-orders-mean

Swenson, A. (2025, January 20). *Trump orders government not to infringe on Americans' speech, calls for censorship investigation.* Ap. https://apnews.com/article/trump-executive-order-censorship-misinformation-social-media-b344d8c75463990ff6e782bc836bb337

Tait, R. (2025, January 31). *Trump's revenge agenda has shocked officials who 'didn't think it was going to be this bad', insiders say.* The Guardian. https://www.theguardian.com/us-news/2025/jan/31/trump-federal-workers-deep-state

Takyar, A. (n.d.). *AI applications in financial compliance: An overview.* LeewayHertz. https://www.leewayhertz.com/ai-in-financial-compliance

Tan, H. (2025, March 7). *Trump, once a crypto skeptic, establishes a bitcoin reserve.* Markets Insider. https://markets.businessinsider.com/news/currencies/donald-trump-strategic-cryptocurrency-reserve-executive-order-bitcoin-price-2025-3

Tantleff, A. & Grimes, H. (2025, January 24). *Cybersecurity executive order — Key implications for the manufacturing industry.* Foley. https://www.foley.com/insights/publications/2025/01/cybersecurity-executive-order-key-implications-for-the-manufacturing-industry

Tether. (2025). Tether reserves and market position: June 2025 report. Tether.to. https://tether.to/reserves-report-june-2025.pdf

The ethical considerations of artificial intelligence. (2023, May 30). *Capitology Blog: Capitol*

Technology University. https://www.captechu.edu/blog/ethical-considerations-of-artificial-intelligence

The impact of social media on political campaigns and elections. (n.d.). Prashikshan IAS. https://www.prashikshanias.com/the-impact-of-social-media-on-political-campaigns-and-elections

The United States of America and Italy sign memorandum of understanding to expand collaboration on countering foreign state information manipulation. (2024, April 17). U.S. Department of State. https://2021-2025.state.gov/the-united-states-of-america-and-italy-sign-memorandum-of-understanding-to-expand-collaboration-on-countering-foreign-state-information-manipulation/

The use of social media in the 2024 US election campaign. (2024, October 25). Media Beats. https://media-beats.com/en/social-media-us-election-campaign-2024

The White House. (2025a, January 20). *Ending the weaponization of the federal government*. https://www.whitehouse.gov/presidential-a.1.1tions/2025/01/ending-the-weaponization-of-the-federal-government

The White House. (2025a, March 6). Fact Sheet: President Donald J. Trump establishes the Strategic Bitcoin Reserve and U.S. Digital Asset Stockpile. WhiteHouse.gov. https://www.whitehouse.gov/briefing-room/statements-releases/2025/03/06/fact-sheet-president-donald-j-trump-establishes-the-strategic-bitcoin-reserve-and-u-s-digital-asset-stockpile/

The White House. (2025b, July 18). Fact Sheet: President Donald J. Trump signs GENIUS Act into law. WhiteHouse.gov. https://www.whitehouse.gov/

briefing-room/statements-releases/2025/07/18/fact-sheet-president-donald-j-trump-signs-genius-act-into-law/

The White House. (2025c, January 23). *Fact sheet: President Donald J. Trump takes action to enhance America's AI leadership*. https://www.whitehouse.gov/fact-sheets/2025/01/fact-sheet-president-donald-j-trump-takes-action-to-enhance-americas-ai-leadership

The White House. (2025d, January 23). *Removing barriers to American leadership in artificial intelligence*. https://www.whitehouse.gov/presidential-actions/2025/01/removing-barriers-to-american-leadership-in-artificial-intelligence

The White House. (2025e, February 23). *Fact sheet: President Donald J. Trump takes action to enhance America's AI leadership*. https://www.whitehouse.gov/fact-sheets/2025/01/fact-sheet-president-donald-j-trump-takes-action-to-enhance-americas-ai-leadership

The White House. (2025f, February 23). *Strengthening American leadership in digital financial technology*. https://www.whitehouse.gov/presidential-actions/2025/01/strengthening-american-leadership-in-digital-financial-technology

The White House. (2025g, March 6). *Establishment of the strategic Bitcoin reserve and United States digital asset stockpile*. https://www.whitehouse.gov/presidential-actions/2025/03/establishment-of-the-strategic-bitcoin-reserveand-united-states-digital-asset-stockpile

The White House (2025b, January 20). *Restoring freedom of speech and ending federal censorship*. https://www.whitehouse.gov/presidential-actions/2025/01/restoring-freedom-of-speech-and-ending-federal-censorship

Title II fast, too furious: The return of net neutrality. (2023, November 1). Brownstein. https://www.bhfs.com/insights/alerts-articles/2023/title-ii-fast-too-furious-the-return-of-net-neutrality-

Torossian, R. (2025, January 16). *President Trump's communications strategy: Possible insight*. O'Dwyer's. https://www.odwyerpr.com/story/public/22407/2025-01-16/president-trumps-communications-strategy-possible-insight.html

Trump on surveillance, protest, and free speech. (n.d.). ACLU. https://www.aclu.org/trump-on-surveillance-protest-and-free-speech

Truth Social. (2025, January 6). eSafety Commissioner. https://www.esafety.gov.au/key-topics/esafety-guide/truth-social

Turk, V. (2025, February 9). *From Dogecoin to $Trump: Everything you need know about the wild world of memecoins*. The Guardian. https://www.theguardian.com/technology/2025/feb/09/from-dogecoin-to-trump-everything-you-need-know-about-the-wild-world-of-meme-coins

U.S. Government Accountability Office. (n.d.). *Protecting personal privacy*. https://www.gao.gov/protecting-personal-privacy

UNESCO. (2018). Media and Information Literacy: Curriculum for Teachers and Trainers. UNESCO.

USA: Trump's attacks on government transparency erode press freedom. (2025, February 2). Reporters Without Borders. https://rsf.org/en/usa-trumps-attacks-government-transparency-erode-press-freedom

Vartabedian, M. (2024, December 17). *AI can take the slog out of compliance work, but executives not ready to fully trust it*. The Wall Street Journal. https://www.wsj.com/articles/

ai-can-take-the-slog-out-of-compliance-work-but-executives-not-ready-to-fully-trust-it-7cd60a16

Wardle, C., & Derakhshan, H. (2017). Information Disorder: Toward an interdisciplinary framework for research and policy making. Council of Europe.

Wike, R. Fagan, M. & Clancy, L. (2024, December 11). *Global elections in 2024: What we learned in a year of political disruption.* Pew Research Center.https://www.pewresearch.org/global/2024/12/11/global-elections-in-2024-what-we-learned-in-a-year-of-political-disruption

Wile, B. (2025, March 7). *First White House crypto summit leaves some investors unhappy*. NBC News. https://www.nbcnews.com/business/personal-finance/white-house-crypto-summit-details-what-to-know-rcna195291

Wilkinson, I. (2025, March 7). *Trump, Stargate, DeepSeek: A new, more unpredictable era for AI?* Chatham House. https://www.chathamhouse.org/2025/02/trump-stargate-deepseek-new-more-unpredictable-era-ai

Wilson, T. & Conlin, M. (2025, February 3). *Exclusive: Trump's meme coin made nearly $100 million in trading fees, as small traders lost money.* Reuters. https://www.reuters.com/markets/currencies/trumps-meme-coin-made-nearly-100-million-trading-fees-small-traders-lost-money-2025-02-03

Zilber, A. (2025, March 7). *Trump signs order to create US strategic crypto reserve: 'Digital Fort Knox.'* New York Post. https://nypost.com/2025/03/07/business/trump-signs-executive-order-establishing-crypto-strategic-reserve

Image References

Banks, C. (2020, November 4). Illustration of a building. [Image]. Unsplash. https://unsplash.com/es/fotos/ilustracion-de-edificio-rojo-y-azul-BY-R0UNRE7w

DeAnda, R. (2020, January 2). Photo of the White House. [Image]. Unsplash. https://unsplash.com/es/fotos/fotografia-de-arquitectura-de-la-casa-blanca-zfKlCKK-Ql0

Feyissa, S. (2025, January 28). A person holding a cell phone. [Image]. Unsplash. https://unsplash.com/es/fotos/una-persona-sosteniendo-un-telefono-celular-en-la-mano-MHgLD0-9VvM

Gauthier, M. (2021, May 7). Person holding a cell phone. [Image]. Unsplash. https://unsplash.com/es/fotos/persona-sosteniendo-un-telefono-inteligente-samsung-android-negro-_kz4Mhc0m_Q

GDJ. (2017, September 7). Vectors. [Image]. Pixabay. https://pixabay.com/es/vectors/corrupci%C3%B3n-enga%C3%B1o-eufemismo-2727571

Kaboompics.com. (2020, November 24). Gold Bitcoin coins and cash in a miniature shopping cart. [Image]. Pexels. https://www.pexels.com/photo/gold-bitcoin-coins-and-cash-in-a-miniature-shopping-cart-5980876

Kanchanara. (2021, September 1). Person using a black tablet. [Image]. Unsplash. https://unsplash.com/es/fotos/persona-usando-una-tableta-negra-9BDBjEl_fDA

Leeloo The First. (2021, July 26). Gold coins on black digital tablet. [Image]. Pexels. https://www.pexels.com/photo/gold-coins-on-black-digital-tablet-8919505

Library of Congress. (2020, January 27). Human, face, clothes. [Image]. Unsplash. https://unsplash.com/es/fotos/el-presidente-donald-trump-jPN_oglAjOU

Miroshnichenko, T. (2020, September 17). Close-up view of system hacking. [Image]. Pexels. https://www.pexels.com/photo/close-up-view-of-system-hacking-5380642

Nilov, M. (2021, July 15). Charts and graphs on paper. [Image]. Pexels. https://www.pexels.com/photo/charts-and-graphs-on-paper-on-a-clipboard-9304917

Norouzi, B. (2021, May 23). Smartphone Samsung Android white background. [Image]. Unsplash. https://unsplash.com/es/fotos/smartphone-samsung-android-negro-con-fondo-blanco-699H4PEEEe0

Omilaev, I. (2023, October 17). A computer chip with the letter A. [Image]. Unsplash. https://unsplash.com/es/fotos/un-chip-de-computadora-con-la-letra-a-en-la-parte-superior-eGGFZ5X2LnA

Pixabay. (2016, March 2). American flag on pole. [Image]. Pexels. https://www.pexels.com/photo/american-flag-on-pole-under-blue-sky-during-daytime-54456

Spiske, M. (2019, March 10). Codes on tilt shift lens. [Image]. Pexels. https://www.pexels.com/photo/codes-on-tilt-shift-lens-2004161/

Traxer. (2021, December 27). A golden coin. [Image]. Unsplash. https://unsplash.com/es/fotos/un-primer-plano-de-una-moneda-de-oro-RYx-dIMb7Lk

Tyson, J. (2021, June 28). Donald Trump and text message from Jon Tyson. [Image]. Unsplash. https://unsplash.com/es/fotos/texto-FgewqOVtwbY

Visuals. (2020, March 26). Man in a black suit with a red heart. [Image]. Unsplash. https://unsplash.com/es/fotos/hombre-en-chaqueta-de-traje-negro-con-corazon-rojo-en-el-cuello-S2wnPueyJx4

Winstead, T. (2021, July 21). A politics text. [Image]. Pexels. https://www.pexels.com/photo/a-politics-text-posted-on-a-globe-8850825

Winstead, T. (2021, July 21). Red and blue notebook. [Image]. Pexels. https://www.pexels.com/photo/red-and-blue-notebook-beside-american-flag-8850737

Made in the USA
Monee, IL
04 November 2026

67379452R00125